TRANSACTIONS

OF THE

AMERICAN PHILOSOPHICAL SOCIETY

HELD AT PHILADELPHIA
FOR PROMOTING USEFUL KNOWLEDGE

NEW SERIES—VOLUME 43, PART 3
1953

STUDIES ON CHROMOSOMES AND NUCLEAR DIVISION

I. Fusion of Nucleoli Independent of Chromosomal Homology.
II. Spontaneous Aberrations, Homologous and Non-homologous Union of Fragments.
III. Pairing, Segregation, and Crossing-over.
IV. Photomicrographs of Living Cells During Meiotic Divisions

L. R. CLEVELAND

Professor of Biology, Harvard University

THE AMERICAN PHILOSOPHICAL SOCIETY
INDEPENDENCE SQUARE
PHILADELPHIA 6

OCTOBER, 1953

Library of Congress Catalog
Card Number: 53–10558

STUDIES ON CHROMOSOMES AND NUCLEAR DIVISION *

L. R. CLEVELAND

CONTENTS

Biology has its prophets.

"The most important discoveries of the laws, methods and progress of nature have nearly always sprung from the examination of the smallest objects which she contains, and from apparently the most insignificant enquiries." Lamarck, *Philosophie Zoologique.*

"I have long been persuaded that the protozoa, when more accurately investigated, will furnish us with cytological data of great biological interest. . . . In a word, I believe that most biological generalizations are still—to use the apt expression of Bacon—mere *anticipationes naturae,* and that the protozoa offer us a means of transforming them into something nearer to a veritable *interpretatio naturae.* . . . The blessed word 'experimental' nowadays covers a multitude of biological sins, yet there is sometimes more virtue in one accurate observation than in ninety-nine inconclusive experiments." Dobell, *Review La Cellule,* 35, 1924.

I. FUSION OF NUCLEOLI INDEPENDENT OF CHROMOSOMAL HOMOLOGY

Nucleoli are of common occurrence in plants and animals and a considerable amount of study has been devoted to them. In general, they disappear in late prophase and reappear in telophase. They are always associated with one or more chromosomes to which they are rather firmly anchored and from which they appear to be produced. The association is not a random one, but is always at a particular point on a chromosome, the more usual position being the ends of chromosomes.

Nucleoli vary in size, and probably in composition, too, since they differ greatly in appearance when observed with phase contrast. Some of the largest nucleoli known—as well as the largest and most favorable chromosomes for cytological studies—are found in certain protozoan flagellates of the order Hypermastigina. There are many genera of this group with only two chromosomes. I have studied the whole life cycle of both nucleoli and chromosomes of several genera and species, but, since all the genera and species studied present practically the same chromosomal and nucleolar behavior, only one species, *Holomastigotoides tusitala,* is considered in this paper. There is also another reason for restricting the account to this species, namely, the fact that the whole life cycle of its chromosomes has been described (Cleveland, 1949a).

This two-chromosome organism must be regarded as a haploid for two reasons: first, one chromosome is considerably longer than the other; second, both chromosomes have terminal nucleoli, and the longer one has also a lateral nucleolus which is located about one-third its length from the centromere end. The centromere of each chromosome may be seen clearly at all times from one cell generation to another, and it is unmistakably plain that both centromeres are terminal. Another unique and valuable feature of this organism for life cycle studies, is the fact that an achromatic figure is always present, and the chromosomes, by means of their centromeres, are always anchored to it (for details of these features, see Cleveland, 1949a text-figs. 3, 4, figs. 1–33). Thus there is never any difficulty in differentiating one chromosome from the other.

After the chromosomes are duplicated in the early prophase, there are, of course, four chromatids present, two short[1] ones, with only terminal nucleoli, and two long ones with both terminal and lateral nucleoli. In very early prophase, the sister chromatids have many gyres of relational coiling, but, as prophase development continues, the relational coiling is gradually lost by unwinding; finally, by late prophase, practically all of it is resolved and each chromatid lies entirely free (fig. 1).

Before beginning with the description of the various types of nucleolar fusion, we should note that there are two races or varieties of *H. tusitala*: in one, the prophase chromatids are single; in the other, they are double, each chromatid being composed of two relationally coiled half chromatids. Each half chromatid, like each chromatid, has the same number of nucleoli as a

* Aided in part by an Institutional Grant from the American Cancer Society.

---

[1] In some of the drawings the short chromosomes actually appear to be as long as the long ones. This results from the fact that in such instances the long chromosome is bent so that a camera lucida tracing of it actually is no longer than the tracing of a short, unbent one. In making the drawings, the camera lucida tracings were followed explicitly. In other words, no correction in length was made for the bending.

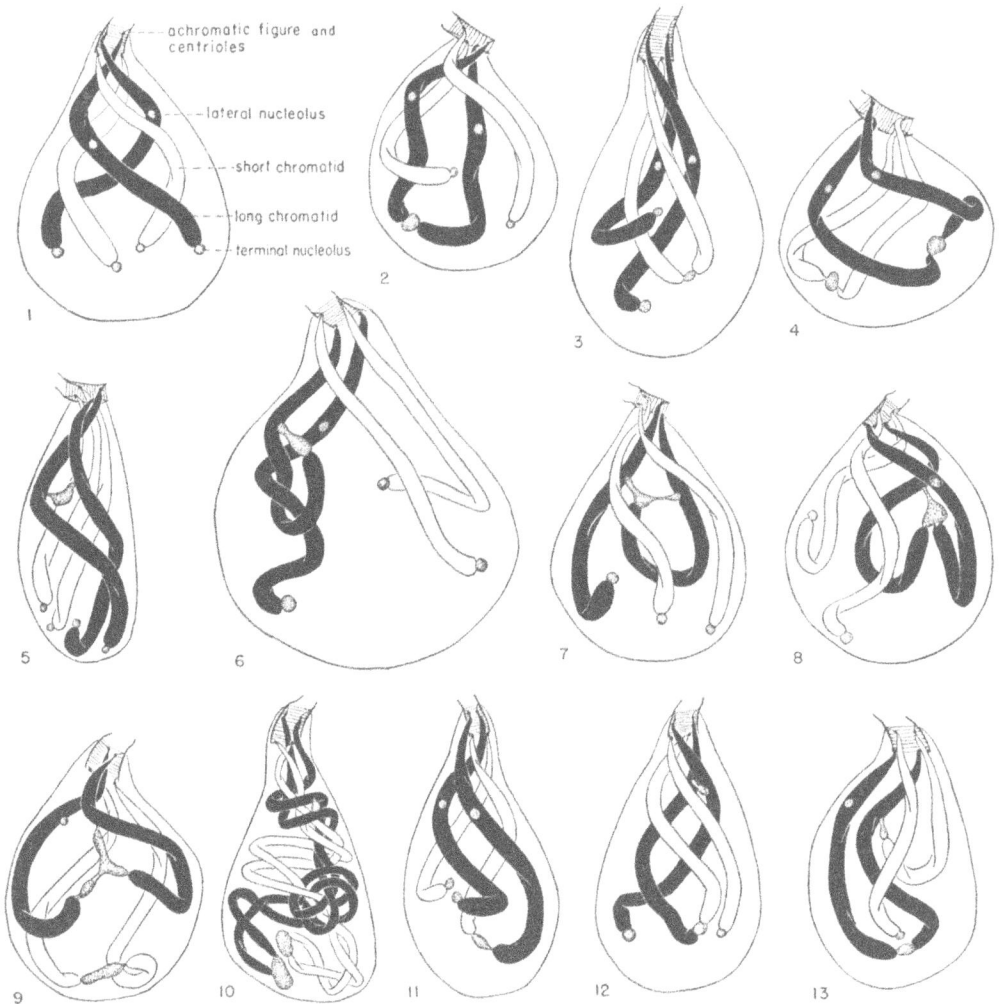

FIGS. 1–13. *Holomastigotoides tusitala.* Prophase chromatids, their nucleoli, and (except for fig. 1) different types of nucleolar fusion. All figures except 10 are fairly late prophases. × 1,400.

1. No fusion of nucleoli.
2. Terminal nucleoli of long chromatids fused.
3. Terminal nucleoli of short chromatids fused.
4. Terminal nucleoli of both short and long chromatids fused.
5. Lateral nucleoli fused.
6. Terminal nucleolus of one long chromatid fused with its own lateral.
7. Terminal nucleolus of one long chromatid fused with both lateral nucleoli.

8. Terminal nucleoli of both long chromatids fused with one lateral.
9. Same as fig. 8 and in addition the terminal nucleoli of the short chromatids are fused.
10. Same as fig. 4 and in addition lateral nucleoli are fused.
11. Terminal nucleoli of a short and a long chromatid fused.
12. Same as fig. 11 and lateral nucleoli are also fused.
13. Terminal nucleoli of long chromatids fused, and one lateral nucleolus fused with a terminal of a short chromatid.

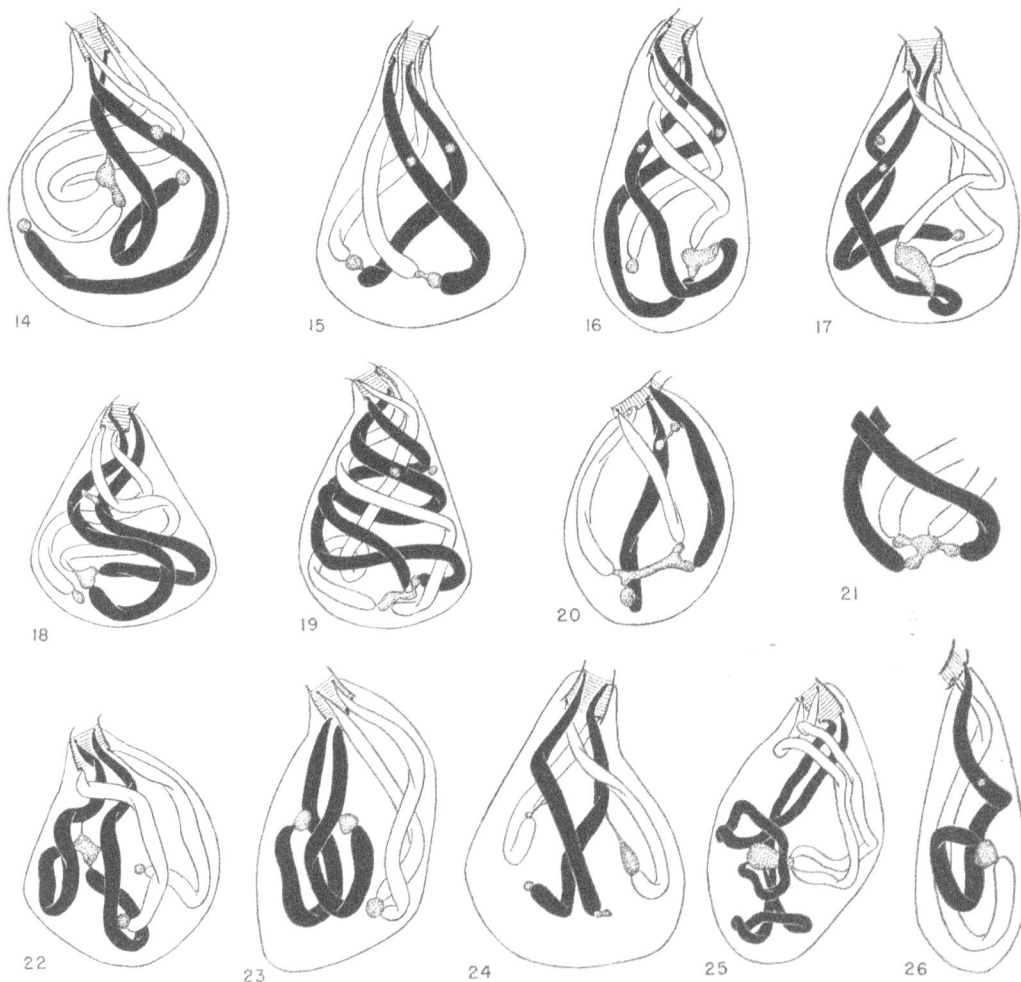

FIGS. 14–26. *Holomastigotoides tusitala.* Different types of nucleolar fusion. Fig. 26 is telophase after cytokinesis; 19 is midprophase; others are fairly late prophases. × 1,425.

14. Terminal nucleoli of short chromatids both fused with a lateral nucleolus.

15. Terminal nucleolus of each short chromatid is fused with a terminal of a long chromatid. No homology whatever.

16. Both terminal nucleoli of long chromatids fused with terminal nucleolus of one short chromatid.

17. Both terminal nucleoli of short chromatids fused with terminal nucleolus of one long chromatid.

18. Same as fig. 17 and in addition lateral nucleoli are fused.

19. All four terminal nucleoli fused.

20. Same as fig. 19 and in addition lateral nucleoli are fused.

21. Same as 20 except only three of the four terminal nucleoli are beginning to pull apart.

22. Terminal and lateral nucleoli of long chromatids have fused to form a common nucleolus.

23. Terminal nucleolus of each long chromatid fused with its own lateral, terminal nucleoli of short chromatids are also fused.

24. Terminal nucleolus of each short chromatid is fused with a lateral nucleolus. No homology whatever.

25. All six nucleoli fused into a single body.

26. Terminal nucleolus of the short chromosome fused with that of the long chromosome.

chromosome. As a matter of fact, it is a chromosome, in morphology and physiology, but not in terminology.

In most instances, the sister nucleoli—terminal as well as lateral—of the half chromatids are fused in prophase. For this reason, and also in order to make the account of nucleolar fusion as simple as possible, all chromatids have been drawn as if they were single, notwithstanding the fact that they are actually double in approximately half the nuclei drawn.

The fact that the nucleoli of sister half chromatids—which always lie close together—are often fused, suggests strongly that fusion occurs when nucleoli come near one another. If this is true, then, there should be more fusion when chromatids are longest. Observations and counts prove this to be correct (see Cleveland, 1949a, figs. 4–19). Fusion of nucleoli, I think, will be found to be so common, when more studies on the entire nucleolar cycle are made, that many organisms, where only one nucleolus has been reported from observations on resting nuclei, will be found to have several nucleoli.

In a majority of nuclei, even in fairly late prophase shortly before nucleoli begin to disappear, there is some fusion, and there is always a certain amount of fusion in earlier prophases. The position and contour of chromatids whose nucleoli are fused are usually altered considerably as a result of fusion (figs. 6–9). Of course, in earlier prophase, when the chromatids appear longer because they have smaller major coils, nucleolar fusion alters their direction only slightly, if at all, because each chromatid, at this time, is several times as long as the greatest diameter of the nucleus.

Since nucleolar duplication occurs at the same time as that of a chromosome, it is therefore difficult to say whether the sister nucleoli fuse immediately after they are formed or whether they merely remain together until some time later, after a considerable amount of the relational coiling of the sister chromatids has come out by unwinding (compare figs. 15b, 16b with 19b, Cleveland, 1949a). But whenever the terminal nucleoli of short and long chromatids are seen as a common body (figs. 11, 12, 15) fusion has undoubtedly occurred. Likewise, when a terminal nucleolus of a long chromatid is seen joined to the lateral of the same chromatid (fig. 6), nucleolar fusion must have occurred. And the same may be said when the terminal nucleolus of a long chromatid is joined to both laterals of the long chromatids (fig. 7), when terminals of both long chromatids are joined to a lateral (figs. 8, 9), when terminals of both short chromatids are joined to a lateral of one long chromatid (fig. 14), when both terminals and both laterals of the long chromatids are joined (fig. 22), when each terminal of the long chromatids is joined to the lateral of the same chromatid (fig. 23), when each terminal of the short chromatids is joined to a lateral (fig. 24), when all four terminals are joined (figs. 20, 21), when all terminals and all laterals are joined (fig. 25), and, finally, when the terminals of the two non-homologous telophase chromosomes are joined (fig. 26).

Observations such as these not only establish the fact that nucleolar fusion occurs, but they also show that it is in complete disregard of chromosomal homology. Studies of living cells show that the bond or linkage between chromatids via their nucleoli is a firm, strong one.

Most examples of nucleolar fusion that have been drawn are fairly late prophases (figs. 10, 19, 20 are exceptions). This has been done purposefully because the reader can see more readily the type of fusion when chromatids appear short (owing to rather tight major coils).

References will be found at the end of the fourth paper of this series.

II. SPONTANEOUS ABERRATIONS, HOMOLOGOUS AND NON-HOMOLOGOUS UNION OF FRAGMENTS

Both spontaneous and induced aberrations or breaks of chromatids have been studied in several organisms, but no one has studied them in an organism such as the flagellate *Holomastigotoides tusitala* with only two chromosomes, each of which may be followed easily throughout its entire life cycle, or from one generation to another. In this organism there is no difficulty at all in determining the precise point where a break occurs on a chromatid because its centromere and distal ends are easily differentiated from each other at all times. It is also possible, except in those rare cases where chromatids break into more than three fragments, to determine the exact point where union occurs, as well as the nature of the fragments that come together when chromatids unite following breaks. Also, the nature of those fragments in which union does not occur may be determined with the same ease and precision.

I have studied aberrations in many genera of hypermastigote flagellates—those with many chromosomes as well as those with only a few—but, since it is impossible, in those organisms with many chromosomes, to determine the precise point where a break occurs, the exact length of the fragment produced, and the specific point where union takes place, it seems desirable to restrict this account to asexual, haploid organisms with two chromosomes where such studies may be carried out. Further, inasmuch as it is easier when the chromosomes are rod-shaped to follow through—as well as to illustrate—breaks and union, the large number of species studied in the genera *Spirotrichonympha* and *Holomastigotoides* with subterminal centromeres also should not be considered. This, then, leaves for consideration two species of *Spirotrichonympha* and one of *Holomastigotoides,* namely *H. tusitala.* The two species of *Spirotrichonympha* do not have achromatic figures to which the centromeres are attached during the whole life cycle of the chromosomes, nor, in either species, does the long chromosome, as in *H. tusitala,* have a lateral nucleolus which readily differentiates it from the short one. For these, and other reasons, *H. tusitala* is the best organism I have seen for a critical study of aberrations.

There is also another highly desirable feature of *H. tusitala* which should be mentioned. In the prophase, the four chromatids are always attached by their centromeres to the four corners of the flat achromatic figure formed by the distal ends of the elongate centrioles which lie at its margins. The centromeres of sister chromatids are invariably attached to opposite corners. For example, those of the two long chromatids are always attached to the opposite upper corners of the achromatic figure (fig. 27) or to the opposite lower cor-

ners (fig. 30). In other words, one never finds one long chromatid attached to the upper right hand corner of the achromatic figure and the other long chromatid attached to the lower left hand corner. This absolutely constant method of attachment of chromatids to the achromatic figure is of great value in determining, for example, whether union to form a dicentric occurs between sister chromatids (fig. 39) or non-sisters (fig. 41).

For more details concerning *H. tusitala* and the whole life cycle of its chromosomes, see Cleveland, 1949*a.* The manner in which nucleolar fusion occurs is considered in the first paper of this series. For an explanation of the technique employed in making the drawings so that any chromatid, or the fragments of any chromatid, may be readily identified, see p. 815, fig. 27.

Aberrations may be seen at any stage in the life cycle of a chromosome, but they are more easily detected in fairly late prophases. It is also easier at this time, when the chromatids appear quite short in contrast to the early prophase, to determine the nature of the centric and acentric fragments that have united following aberrations. Therefore, most of the illustrations are of aberrations of prophase chromatids.

But when did the breaks which one can see best in the late prophase occur? This is a question that cannot be answered definitely without more study. However, I have the feeling that they occur in early prophase, shortly after chromosomal duplication. At this time there is very little union; most fragments have free ends (fig. 34). Between this stage and the fairly late prophase—which incidentally is also the resting stage in this organism—is probably the time when most fragments undergo union. Of course, one could argue that chromatids break in the late prophase and since they are resting at this stage there is ample time for union to occur. There is, however, another fact which suggests this is probably not the case. In most preparations one can find a few aberrations, but they are always more abundant in those where there is a good deal of reproduction. It is thus difficult to get away from the feeling that whatever causes these aberrations is in some way rather intimately connected with the duplication phase of the chromosome cycle.

Certainly most, if not all, of the aberrations shown are lethal. The question then arises, how long can a cell live after such aberrations occur? Again the answer is somewhat indefinite, but some cells certainly survive through the prophase resting period and through the mitosis that follows because aberrations may be found after cytokinesis (figs. 51, 52) which aberrations must have occurred in prophase, either before or during the resting period.

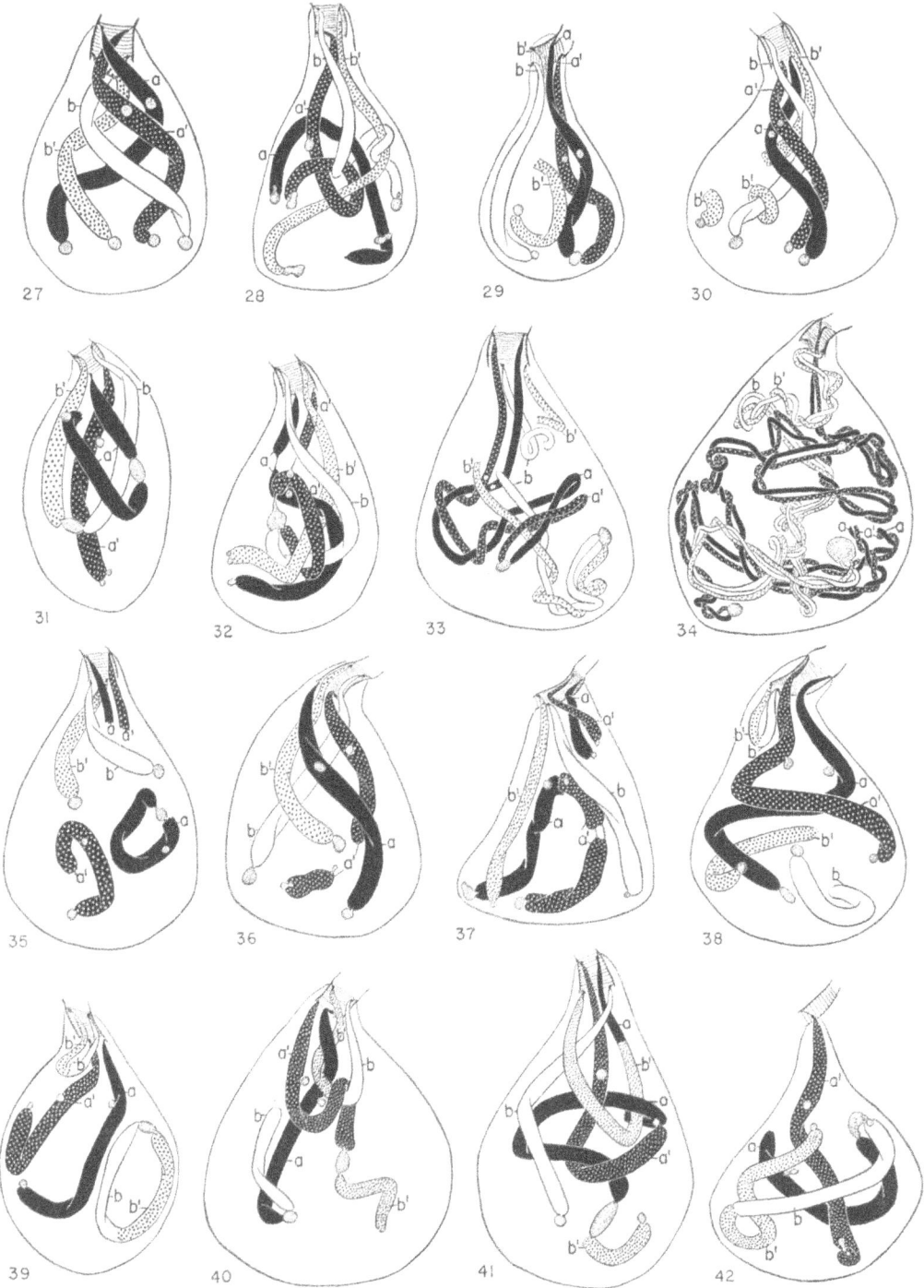

An analysis of the large number of aberrations observed—only a few of which have been drawn—shows that a chromatid may break at any point along its entire length. However, there are fewer aberrations near the distal ends than elsewhere.

The free ends of non-homologous chromatids unite just as often as those of the homologous ones. This leads one to suspect that union depends on chance meeting of free ends rather than the attraction of one free end to another. This idea is further substantiated by the fact that unbroken, homologous chromatids of *H. tusitala* do not pair at any stage in their life cycle. However, it should be noted that in several species of the related genus *Spirotrichonympha* there is pairing, in two-chromosome, haploid organisms, of homologous chromatids from one end to the other. For this reason alone such organisms are less desirable than *H. tusitala* for studying union of fragments.

Since *H. tusitala* has such a high rate of lethal aberrations, one cannot help but wonder whether there are

---

FIGS. 27–52. *Holomastigotoides tusitala.* Figs. 28–50 show how spontaneous aberrations of chromatids produce many types of centric and acentric fragments, how some of these fragments remain free, how others, without regard to homology, unite by union of their free ends, and how some are united or joined by fusion of their nucleoli. Figs. 33 and 34 are fairly early prophases; 51, 52 are telophases; all others are fairly late prophases. Only distal ends of centrioles are shown. They are the heavy lines at the margins of the achromatic figure (right and left margins in all figures except 29, 37, and 48 where they are up and down). × 1,450.

27. This figure shows normal unbroken chromatids and also explains the drawing technique used in those figures (28–52) where aberrations occur. Two chromosomes are present, a long one and a short one, each of which is composed of two sister chromatids. All chromatids have a centromere and a terminal nucleolus, and each chromatid of the long chromosome has in addition a lateral nucleolus. One chromatid of the long chromosome is drawn solid black and is labeled *a*, while its sister is drawn black with white stipples and is labeled *a'*. One chromatid of the short chromosome is drawn white and is labeled *b*, while its sister is drawn white with black stipples and is labeled *b'*. When a chromatid is broken, its fragments are all drawn and labeled in the same way.

28. Long chromatid *a* broke at centromere end.

29. Short chromatid *b'* broke a short distance posterior to its centromere, producing a centric and an acentric fragment.

30. Short chromatid *b'* broke once near its mid-region and again between this point and the terminal nucleolus, producing three fragments, one of which is centric and remains connected to the achromatic figure, and two which are free. The ends of the free fragment without a nucleolus have united to form a ring which encircles short chromatid *b* a short distance from its terminal nucleolus.

31. Long chromatid *a* broke near its lateral nucleolus, producing a centric and an acentric fragment. Terminal nucleolus of acentric fragment fused with the free (non-centromere) end of the centric fragment, thus inverting more than half of this chromatid. Other end of this chromatid lies free and retains its lateral nucleolus, which now is almost terminal. Long chromatid *a'* has two terminal nucleoli because this nucleus is the double-chromatid variety in which the terminal nucleoli of half chromatids sometimes are not fused.

32. Long chromatid *a* has almost broken in two slightly anterior to its lateral nucleolus, long chromatid *a'* has broken in two at the same point, and the free end (not nucleolar end) of its acentric fragment has united with the terminal nucleoli of this same fragment and that of unbroken short chromatid *b*. This nucleus is of the double-chromatid variety and the terminal nucleoli of the half chromatids of short chromatid *b'* are not fused.

33. Both short chromatids *b* and *b'* broke about one-third of the way from their centromere ends, and all four ends lie free. Terminal nucleoli of short and those of long chromatids fused.

34. Both long chromatids *a* and *a'* are broken slightly posterior to their mid-regions, leaving four free ends (to the right of the large nucleolus formed by the fusion of the terminal nucleoli of the short chromatids *b* and *b'*). Terminal nucleoli of long chromatids fused; so are their laterals. Since this is a fairly early prophase, a considerable amount of relational coiling of chromatids remains.

35. Long chromatids *a* and *a'* broke mid-way between their lateral nucleoli and their centromere ends, producing two centric and two acentric fragments. No union of any fragments except the free end of acentric fragment *a* has united with the terminal nucleolus of this fragment (lower right).

36. Long chromatid *a'* broke fairly near its terminal nucleolus, leaving a centric fragment with a lateral nucleolus and an acentric fragment with a nucleolus on each end (this fragment probably formed a ring earlier, as in fig. 35, which in opening up left nucleolar material on both ends). There is no evidence that a free end can form a nucleolus. Nucleoli are seen on such ends only when the break occurs either in or very near a lateral nucleolus.

37. Both long chromatids broke in the region of their lateral nucleoli, and the acentric fragment of *a'* is either almost broken in two again near its mid-region or is rejoining at this point. The two centric fragments have developed nucleoli at their free ends (these are lateral nucleoli which, because of the breaks, have become terminal).

38. Both short chromatids broke fairly near their centromere ends. The centric fragments have united to form a dicentric, while the acentric fragments lie free.

39. Same as previous figure except acentric fragments of short chromatids have fused at both ends, forming a ring. However, the ring at one end is only held together by the fusion of the terminal nucleoli of these fragments.

40. Both short chromatids broke near their mid-regions and the free ends of the centric fragments united to form a dicentric. The terminal nucleolus of acentric fragment *b* has fused with the terminal nucleolus of long chromatid *a*, and the terminal nucleolus of acentric fragment *b'* has fused with the terminal nucleolus of long chromatid *a'*. Both long chromatids pass through the loop or ring formed by the dicentric.

41. Long chromatid *a* broke between its centromere and its lateral nucleolus (nearer the nucleolus), and short chromatid *b'* broke not very far from its terminal nucleolus. Then the free ends of long chromatid *a* and short chromatid *b'* united to form a long, looped dicentric composed of non-homologous chromatids (more of *b'* than *a*). The acentric fragment of long chromatid *a* is joined: (1) by its terminal nucleolus to the non-nucleolar end of the acentric fragment of short chromatid *b'* and (2) by its lateral nucleolus to terminal nucleolus of unbroken long chromatid *a'*. Note acentric fragment of long chromatid *a* passes through loop of dicentric after which it passes first under and then over long chromatid *a'*.

42. Short chromatids *b* and *b'* and long chromatid *a* all broke at or very near their centromere ends. Short chromatids have united at their non-nucleolar ends, and the terminal nucleolus of short chromatid *b* has fused with the terminal of long chromatid *a*.

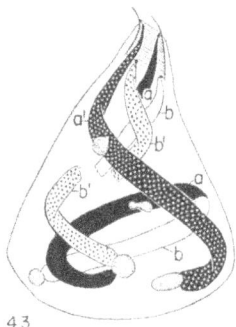

43

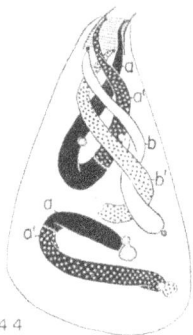

44

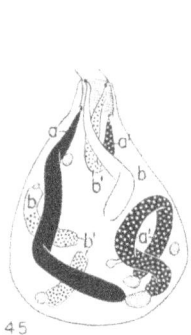

45

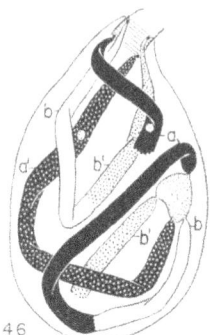

46

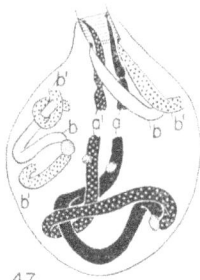

47

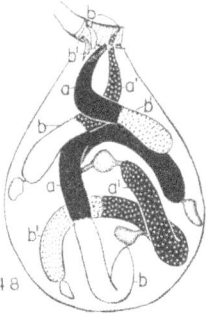

48

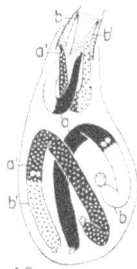

49

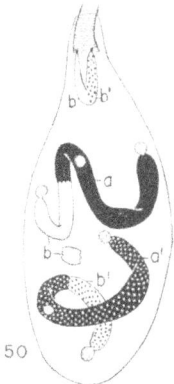

50

51

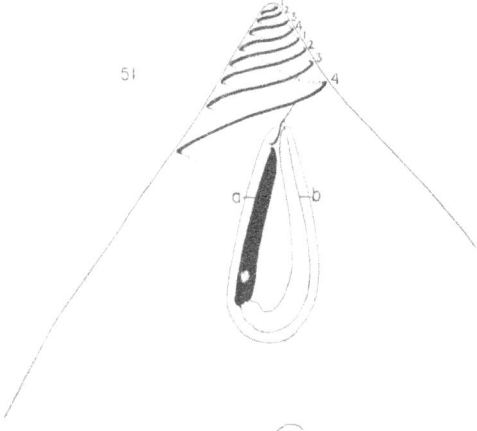

52

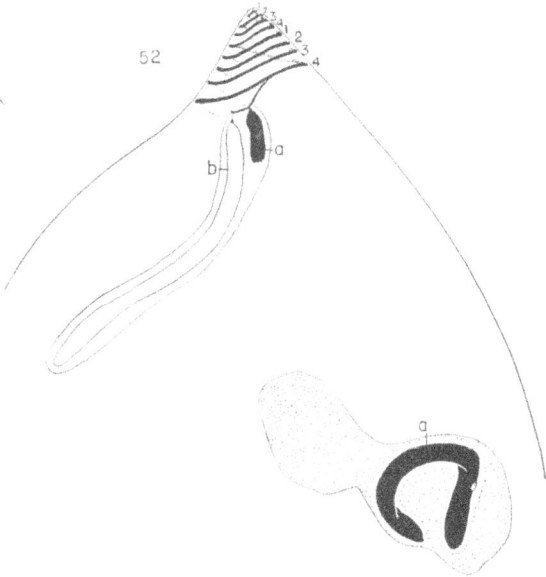

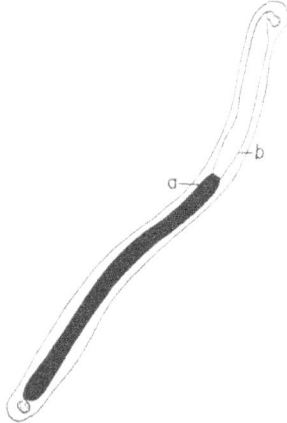

not also some sublethal but undetectable ones in which a small section or region of one chromatid is transferred to another, or in which a region undergoes inversion. Small translocations and inversions, if they occur, may be a means of promoting evolution in asexual haploid organisms. We certainly need an explanation for the production of so many species in such organisms; species, which, from any biological or taxonomic point of view one wishes to take, are as valid as in those forms where sexual reproduction occurs.

References will be found at the end of the fourth paper of this series.

FIGS. 43–52. *Holomastigotoides tusitala.* For explanation see page 815, figs. 27–52. × 1,450.

43. Long chromatid *a* broke mid-way between its centromere and lateral nucleolus. Short chromatids *b* and *b'* each broke at its mid-region. Centric fragment of long chromatid *a* is free, while the acentric fragment of this chromatid has united with the acentric fragment of short chromatid *b*. Centric fragments of short chromatids *b* and *b'* are free, and the terminal nucleolus of acentric fragment *b'* has fused with the terminal nucleolus of the acentric fragment *a*.

44. Long chromatids *a* and *a'* broke slightly posterior to their lateral nucleoli and the broken ends of their centric fragments united to form a dicentric. Acentric fragments of chromatids *a* and *a'* have joined at their non-nucleolar ends. Short chromatid *b'* broke near its terminal nucleolus and this short fragment, by means of its terminal nucleolus, has fused with the terminal nucleolus of long chromatid fragment *a'*.

45. Long chromatid *a'* broke mid-way between centromere and lateral nucleolus; short chromatid *b'* broke near its mid-region; free ends of the centric fragments of these chromatids united to form a dicentric, which, in this case, is anchored by both centromeres to the same centriole (upper and lower corners of achromatic figure). Fragment of long chromatid *a'* is joined by its terminal nucleolus to the terminal nucleolus of unbroken long chromatid *a*. Fragment of short chromatid *b*, which has a nucleolus at one end, has joined one of the fragments of *b'*; other fragment of *b'* lies free (has terminal nucleolus on its lower end).

46. Short chromatids *b* and *b'* broke about two-thirds of the distance from their centromere ends and the free ends of their centric fragments united to form a dicentric. Long chromatid *a* broke just posterior to its lateral nucleolus and the free end of its acentric fragment united with the free or non-nucleolar end of the acentric fragment of short chromatid *b*, while its end with a terminal nucleolus joined, by nucleolar fusion, with the nucleolar end of unbroken long chromatid *a'* and also with the nucleolar ends of fragments of short chromatids *b* and *b'*. Two free ends: one of acentric fragment *b'* (near lower margin of nucleus) and one of the centric fragment of long chromatid *a* (just posterior to its lateral nucleolus).

47. All four chromatids broke about one-third distance from their centromere ends, long ones broke slightly nearer centromeres, thus allowing their lateral nucleoli to go with their acentric fragments. Broken ends of centric fragments of long chromatids *a* and *a'* and of short chromatids *b* and *b'*, all four of them, are free; nucleoli of distal ends of acentric fragments of long chromatids are fused. No fusion of free ends of acentric fragments of short chromatids; two fragments of *b'*; one fragment of *b*; terminal nucleoli of fragments *b* and *b'* fused.

48. Long chromatids *a* and *a'* broke between centromeres and lateral nucleoli (nearer the latter). The short chromatids *b* and *b'* each broke twice, once near centromere and again in their mid-region. One acentric fragment of *b* united with the centric fragment of *a'* and one acentric fragment of *b'*, the one with a terminal nucleolus, united with *a'*. The other acentric fragments of *b* and *b'* united with the acentric fragments of *a* and *a'*, *b* with *a* and *b'* with *a'*. Lateral nucleolus of *a'* fused with terminal of *a*; lateral nucleolus of *a* joined the free end of *b'*; terminal nucleoli are on distal ends of *a'* (right), *b* (lower left), and *b'* (upper left).

49. All four chromatids broke about one-third of distance from their centromere ends. Then broken ends of centric fragments of long chromatid *a* and short chromatid *b* united to form a dicentric; broken ends of centric fragments of long chromatid *a'* and short chromatid *b'* united to form a dicentric. No homology in the unions which formed the double dicentric. Broken end of acentric fragment of long chromatid *a* united with that of short chromatid *b*, leaving nucleolar ends of both chromatids free; broken end of acentric fragment of long chromatid *a'* united with that of short chromatid *b'*, leaving nucleolar ends of both chromatids free.

50. Long chromatids *a* and *a'* both broke at or very near their centromeres; short chromatids *b* and *b'* both broke about one-third distance from their centromere ends, and their free ends of their centric fragments united to form a dicentric. The free end of acentric fragment of short chromatid *b* united with free (non-nucleolar) end of long chromatid *a*. The free end of acentric fragment of short chromatid *b'* united with free (non-nucleolar) end of long chromatid *a'*. Ununited small fragment which may be from either chromatid *b* or *b'*.

51. Telophase after cytokinesis. This is an example where, in the prophase, one long chromatid (either *a* or *a'*) broke slightly posterior to its lateral nucleolus, one short chromatid (either *b* or *b'*) broke near its mid-region, and the free ends of the centric fragments of these chromatids (which are now chromosomes) united to form a dicentric. Since both chromatids at this time and during nuclear division were anchored to the same centriole by their centromeres, the dicentric was not pulled in two. Acentric fragments of these same chromatids (now chromosomes) united by their free or non-nucleolar ends and in the process of reorganizing a new nuclear membrane from portions of the old one, they remained in a common portion of the new nuclear membrane. However, as in the previous figure, that portion of the new nuclear membrane, which, in this figure, contains two acentric fragments, has not united with the portion containing two centric fragments to form a common new nuclear membrane.

52. Telophase after cytokinesis. This aberration probably occurred in prophase, at which time either long chromatid *a* or *a'* (which is now a chromosome) broke about mid-way between its centromere and lateral nucleolus. There was no union following the break, so that the acentric fragment, surrounded by its portion of the reorganized nuclear membrane, became lost in the cytoplasm (lost its connection with the achromatic figure and hence became separated from the other reorganized portion of the nuclear membrane. Under normal conditions the two portions of nuclear membrane would be fused by this time to form a common nuclear membrane). The centric fragment of long chromatid *a* maintained its connection with the achromatic figure during nuclear division, and in late telophase its portion of the new nuclear membrane fused with that of short chromatid *b*.

III. PAIRING, SEGREGATION, AND CROSSING–OVER

CONTENTS

INTRODUCTION

Pairing, segregation, and crossing-over have all been studied for a long time, by many cytologists from many points of view. Indeed, the number of papers that have been written on these subjects is so great that merely a listing of them would be a long paper in itself. The reader must wonder then—and so do I sometimes—why I wish to extend such a list.

Nothwithstanding the exhaustive studies these subjects have received, there is considerable disagreement on certain questions and it is an easy matter to ask basic questions regarding pairing and crossing-over which cannot be answered—and this will still be true after I have finished. However, I do feel I have found some material in certain organisms—which hitherto have not been studied from this point of view—that will help toward an elucidation of certain points here and there, both in the unknown and in the disputed or confused areas of the picture.

Pairing, segregation, and crossing-over are so interrelated that it is impossible to understand certain features of one without considering the others. However, in describing them one must adhere to some sort of order, for only one can be considered at a time. Since crossing-over, in my opinion, is directly dependent upon the kind of pairing that occurs, it is obviously undesirable to consider this subject first. Likewise, whether there is segregation of sister chromatids or not during diploid mitoses depends on the nature of the pairing that precedes poleward movement. Thus, pairing should be considered before segregation.

A detailed review of the literature is unnecessary since text-books such as Wilson (1925), Sharp (1934), Darlington (1937), White (1945), De Robertis, Nowiniski, and Saez (1948)—to mention only a few of those published in English—give fairly complete accounts of pairing, segregation, and crossing-over, as well as further references to these subjects.

MATERIAL

Two organisms have been used in most of the observations reported here. They belong to the genera *Holomastigotoides* and *Barbulanympha*. Several other genera have been studied in as much detail as these but, inasmuch as they show very little these do not show, it seems unnecessary to devote time and space to a consideration of them. They do serve, however, to indicate that the observations reported here are not restricted to a few organisms, but to many of those that undergo mitosis and meiosis.

Several species of *Holomastigotoides* have been described, although the descriptions, except for two species (Cleveland, 1949a), are very brief. There are also a large number of undescribed species that I have studied in detail (and may describe if I ever find time). This account, however, deals with only one species, *H. diversa*, in which cytoplasmic division is always longitudinal and nuclear division is always mitotic. This species consists of both haploid ($n = 2$) and diploid ($2n = 4$) varieties or races, and there is never any transition from one variety to the other; both invariably lead an independent existence. Further, each haploid and each diploid variety is always subdivided into two varieties or races which differ from each other in that one invariably has single-chromatid chromosomes and the other double-chromatid chromosomes. Thus, in this species there are four distinct varieties which may be distinguished from one another quickly and easily in any stage of development (figs. 53–60). Instead of being a puzzle—as it was to me before I understood it and as it will be to the reader, too, if he does not take time to get an understanding of the chromosome situation in each variety before going on with the paper— these varieties present an ideal condition for studying pairing, segregation, and crossing-over. The chromosomes are of diagrammatic clarity and each chromosome, chromatid, and half chromatid may be followed through its whole life cycle. The varieties of this organism also illustrate so very clearly the precise differences in segregation in mitosis and in meiosis, as well as how meiosis so very definitely increases the possibility of crossing-over.

Four species of *Barbulanympha* have been described (Cleveland *et al.* 1934), but only one is considered here, since it is sufficient to make clear the differences between mitosis and meiosis in segregation and crossing-over. A more detailed account of meiosis in *Barbulanympha* is being prepared and will appear elsewhere later. Also, in the fourth paper of this series there are several photomicrographs of living cells of this organism undergoing meiosis.

MITOTIC PAIRING [1]

GENERAL REMARKS

Most of the observations that have been made on pairing have been on meiotic chromosomes and, of course, in diploid organisms. The few that have been made on mitotic chromosomes have also been in diploid organisms. No one, so far as I know, has made a study of the pairing of chromatids in haploid, asexual organisms. Yet, in flagellates alone, there is a wealth of such material.

In diploid organisms very few investigators have even considered the possibility of chromatid pairing, either during mitosis or meiosis; yet such pairing is not rare in mitosis and is perhaps universal in two-division meioses. Attention has been focused too sharply on chromosomes. Also there are no organisms—besides the hundreds of species of hypermastigote flagellates—where chromatids may be followed easily and clearly through their whole life cycle (they may be seen at certain stages in some). Hence, it is not surprising to find chromatid pairing has received so little attention.

I must admit that when I first studied nuclear division in haploid hypermastigotes I thought what I now know to be the pairing of sister chromatids was either an undivided chromosome or one in the process of division. Later, when I really saw and understood the manner in which chromosomes duplicate themselves, then I realized the chromatids were pairing, and in the same way that chromosomes are supposed to pair. Actually, if chromosomes pair, why cannot chromatids and half chromatids do the same, because there is no real structural difference (if centromeres are left out of consideration) between a chromatid—or even a half chromatid—and a chromosome. The difference is purely an artificial one that results from the retention of convenient terms that have no relation to reality (we must not allow terminology, however convenient it may be, to mislead us).

If one stops to think about it, there is even more reason for sister chromatids to pair than for homologous chromosomes, because they have the same homology as chromosomes and are closer together in time than chromosomes. And of course the same applies to the pairing of half chromatids whenever they are present.

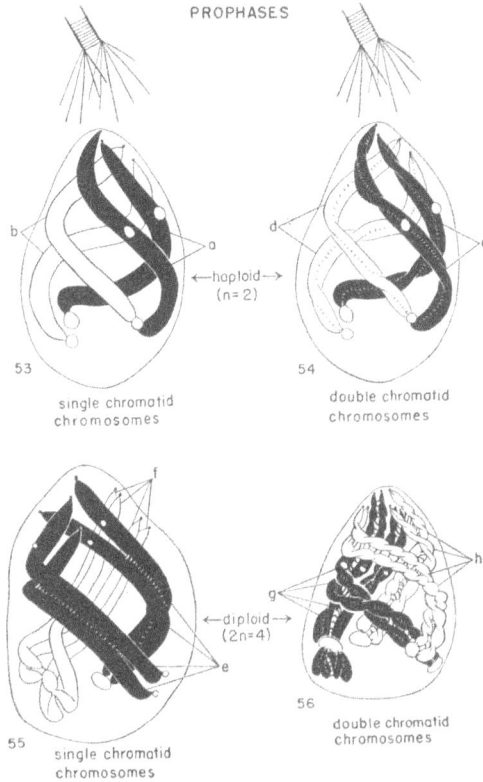

FIGS. 53–56. All are fairly late prophases of *Holomastigotoides diversa*. Note two types or races of haploids (53, 54) and two of diploids (55, 56) present in this species. Each chromatid, regardless of whether it is single or double, has a terminal centromere. Also each chromatid, as well as each half chromatid, has a terminal nucleolus, although fusion of nucleoli is common. In addition, each long chromatid, as well as each long half chromatid, has a lateral nucleolus. × 1,150.

53. The two single chromatids, produced by the duplication of the long chromosome, are dark (*a*) and the two produced by the short chromosome are light (*b*).

54. Same as fig. 53 except each of the long, half-sister chromatids (*c*) and each of the short half-sister chromatids (*d*) is composed of two sister half chromatids. Also note pairing of the sister half chromatids.

55. Same as fig. 53 except this nucleus is diploid and there is pairing in the homologous, long, sister chromatids (*e*) and the homologous, short, sister chromatids (*f*).

56. Same as fig. 55 except each chromosome is composed of two half-sister chromatids, instead of two sister chromatids as in 55, and each half-sister chromatid in turn is composed of two relationally coiled sister half chromatids. However, only two homologies are present (*h* and *g*).

[1] The mitotic pairing described here is intimate. It is not the so-called "somatic pairing," but is comparable with that which usually occurs at meiosis.

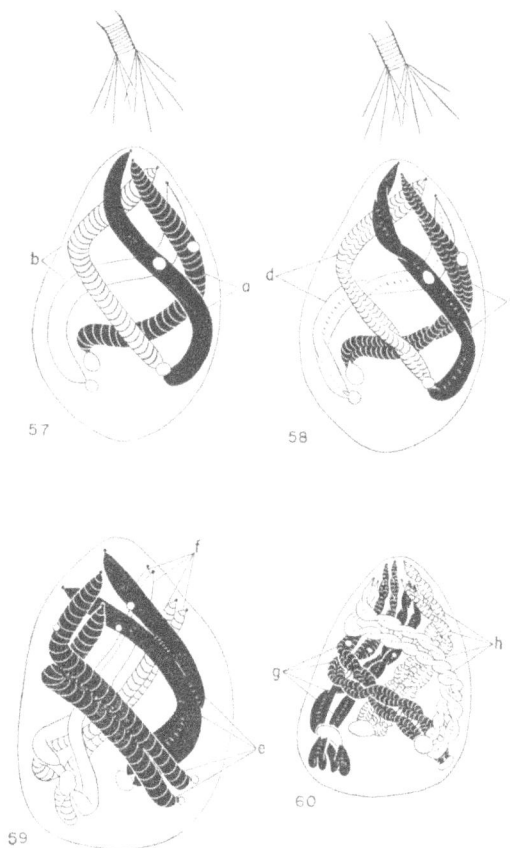

FIGS. 57–60. The four races of *Holomastigotoides diversa*. All are fairly late prophase. Explanation of technique employed in figs. 61–141. In the haploids with single chromatids, such as fig. 57, the sister chromatids are not drawn alike (they differ in degrees of darkness and lightness). Likewise, in haploids where the chromatids are double, such as fig. 58, the half-sister chromatids are not drawn alike, while the sister half chromatids are drawn alike. The presence of a common centromere and many gyres of relational coiling throughout mitosis in the sister half chromatids always serves to differentiate one from the other. In the diploids with single chromatids, such as fig. 59, the sister chromatids of one homologous chromosome are drawn differently from those of the other, while the sisters in each chromosome are drawn in the same manner. In the diploids with double chromatids, such as fig. 60, the half-sister chromatids and the sister half chromatids are drawn alike, but the relational coiling between sister half chromatids, which remains throughout nuclear division, readily distinguishes them from the sister half chromatids of the previous generation (which are now half-sister chromatids). × 1,150.

57. Haploid ($n = 2$). *a.* Two single chromatids derived from the long chromosome; *b.* two single chromatids derived from the short chromosome.

58. Haploid ($n = 2$). *c.* Two double chromatids derived from the long chromosome; *d.* two double chromatids derived from the short chromosome.

In the reorganization of the nuclear membrane the manner in which an old membrane is lost and a new one is produced in hypermastigotes during nuclear division is so intimately related in some of these flagellates to the pairing of chromatids and chromosomes—and this in turn to segregation in diploids—that it must be considered briefly along with pairing. However, a more detailed consideration of this interesting process, which varies considerably in different genera, must be left for a separate paper.

1. HAPLOIDS WITH SINGLE CHROMATIDS

a. Pairing of Sister Chromatids

The formation of the new nuclear membrane begins by a separation of the nuclear contents into granular and hyaline regions. The granular region, which is usually the larger, is eventually completely discarded into the cytoplasm where it may be followed for some time. If the hyaline region, for example in a two-chromosome haploid organism, forms a separate envelope or membrane around each of the four chromatids (fig. 65*a*) before it is time for pairing to begin, pairing does not occur because the enclosure of each of the homologous sister chromatids in a separate portion of the developing new nuclear membrane makes pairing impossible (fig. 68*a*). Perhaps the situation will be clearer if stated in the opposite way, namely, if pairing of sister chromatids does not begin before the portions of the new nuclear membrane are formed, it is prevented by their formation. On the other hand, if pairing begins slightly earlier, or if the formation of these new portions of the nuclear membrane is delayed somewhat, then sister chromatids pair (figs. 66*a*, 67*a*, 69*a*, 72*a*). However, only those regions of them which are in a common portion of the membrane pair. Sometimes, for example, the short sister chromatids are both in separate membranes and lie far apart while at the same time either small or fairly large portions of the long sister chromatids are in a common membrane (figs. 66*a*, 67*a*, 70*a*, 72*a*). But almost as often the situation may be the reverse, namely, the long chromatids may lie far apart with each in a separate membrane, while the short chromatids are paired (fig. 69*a*).

Fusion of terminal nucleoli sometimes appears to play a role in keeping the sister chromatids together, particularly their distal ends, while a common membrane is formed around all or a part of them (figs. 78*a*, 79*a*).

It is rare to see pairing from one end to the other in either the short or the long chromatids. Pairing of some portion, in either the shorts or the longs, is the

59. Diploid ($2n = 4$). *e.* Four single chromatids, two derived from each of the homologous long chromosomes; *f.* four single chromatids, two derived from each of the homologous short chromosomes.

60. Diploid ($2n = 4$). *g.* Four double chromatids, two derived from each of the homologous long chromosomes; *h.* four double chromatids, two derived from each of the homologous short chromosomes.

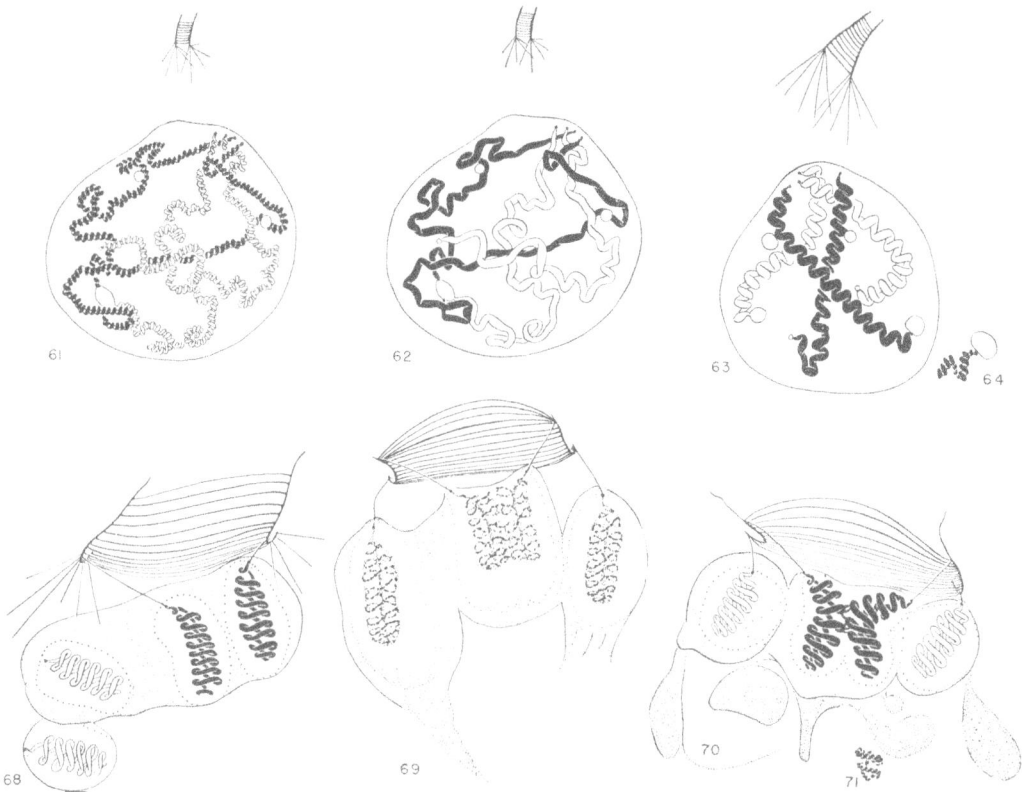

Figs. 61–64, 68–71. *Holomastigotoides diversa*. Haploid ($n = 2$). Single chromatid chromosomes. Prophase to anaphase.
× 1,200.

61. Prophase some time after duplication of chromosomes. Two long (dark) and two short (light) chromatids with minor coils in each and the beginning of major coils. The four centromeres lie fairly close together but have no connection with the early achromatic figure which lies some distance anterior to the nucleus. Three terminal nucleoli fused.

62. Same as fig. 61 except that minor coils are omitted in order to show more clearly the origin of major coils.

63. Later prophase. Only major coils shown. No fusion of nucleoli. Achromatic figure is larger but still lies some distance away from nucleus and has made no connection with the centromeres.

64. Detail of a portion of one chromatid of fig. 63 to show how minor coils are incorporated in the majors.

68. Late prophase or metaphase. Note major coils are now larger and fewer in number in each chromatid. Achromatic figure much larger, is near nucleus, and the two long (dark) chromatids are connected with it, the connection being made by a free astral ray joining a centromere. A small portion of the nuclear membrane is being cut off around each chromatid, and hence there is no pairing.

69. In this figure, which is slightly later than the previous one (all four chromatids are connected with ends of central spindle), no attempt has been made to show either major or minor coils plainly; instead, the matrix of each chromatid and the fine projections from the chromonemata are shown. Note particularly a straight, central rod which extends from the centromere to the distal end of each chromatid. The two short chromatids in the centre are intimately paired except for a small region at their centromere ends. Their matrices are so intimately connected that one can only see where they meet. Their chromonemata are also so closely connected that the folded, figure 8-like major coils (see other figures where only major coils are drawn) of one chromatid seem to extend into those of the other. Note clear area around centromeres. Nuclear membrane is breaking up. Free astral rays omitted.

70. Slightly later stage which may be considered an anaphase. Emphasis is placed on the major coils and the breaking up of the nuclear membrane. The two short chromatids (light) lie far apart and are in almost completely separated portions of the developing new nuclear membrane, while the long (dark) ones, for the most part, lie in a common portion and are slightly paired in the centre (note how firmly some major gyres of one chromatid are connected to those of the other). Free astral rays omitted.

71. Detail of a portion of a chromatid of fig. 70 to show how minor coils are incorporated in majors.

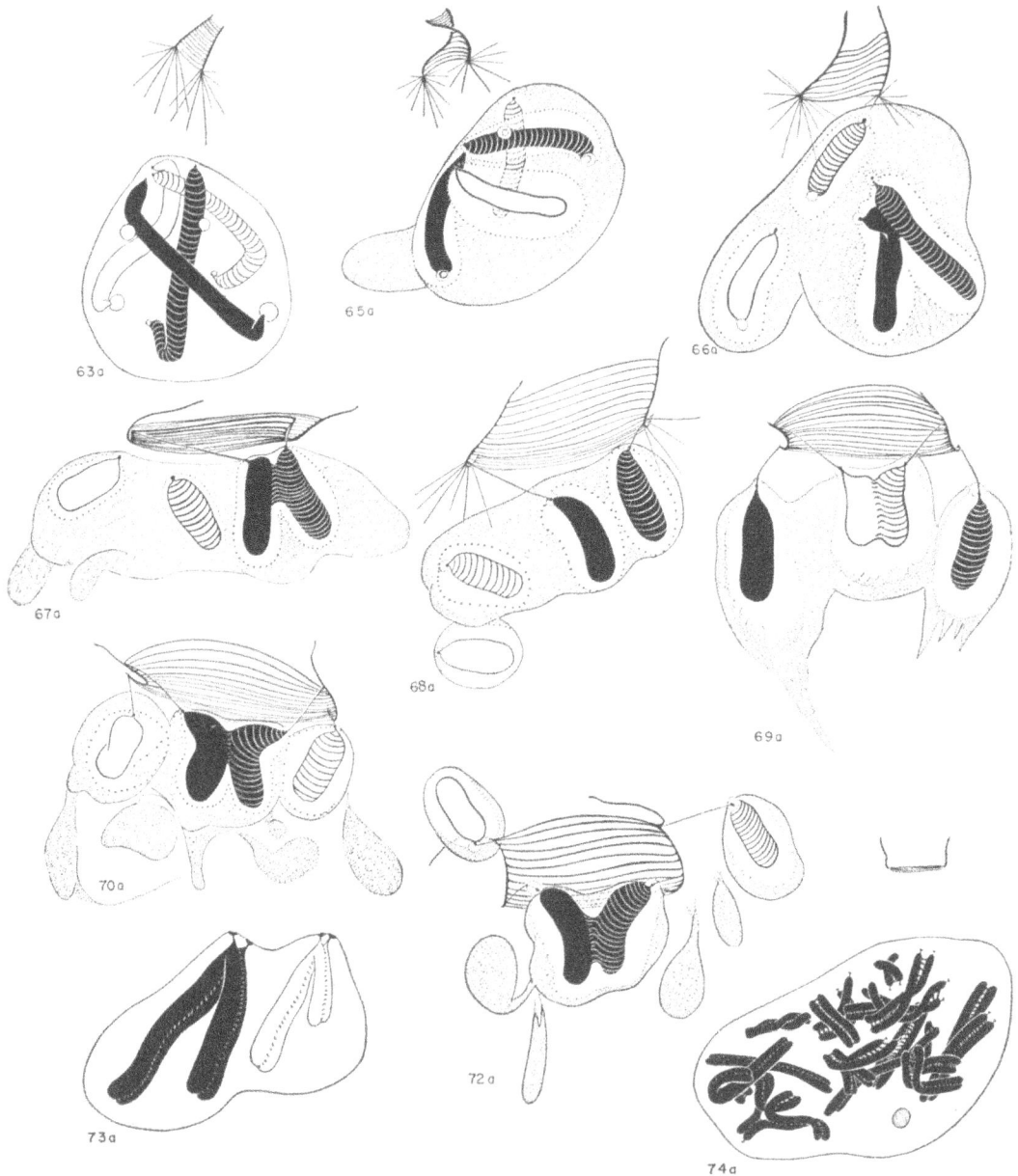

FIGS. 63a, 65a–70a, 72a. *Holomastigotoides diversa*. Haploid ($n = 2$). Single chromatid chromosomes. Prophase to anaphase. Figs. 63a, 68a, 70a same as 63, 68, and 70. Fig. 73a is *Spirotrichonympha* and 74a is *Trichonympha*. Major coils omitted in all chromatids in order to show pairing better. $\times$ 1,150.

63a. Relational coiling of sister chromatids (long ones, dark; short ones, light) is practically all out. Pairing has not begun; nor has the formation of a portion of a new nuclear membrane around each chromatid.

65a. Later stage. Chromatids appear shorter (because major coils are larger in diameter and fewer in number). The clear area around each chromatid represents that portion of the parent or old nuclear membrane which will survive and from

usual picture. One also seldom sees pairing for any distance in both short and long chromatids in the same nucleus. In all metaphases and anaphases where one set of sister chromatids is paired and the other is not—such as shown in figures 69a, 70a, 72a—the set of paired chromatids is always in the centre, and the paired set of chromatids invariably has longer chromosomal fibres (a stickler I suppose might say chromatidal fibres). This means that the paired regions of the chromatids are held together with sufficient firmness that the elongating central spindle portion of the achromatic figure must exert force on them before they will move apart. In other words, the increase in length of the central spindle pulls the chromatids first to the centre of the nucleus and then apart by the tension it places on the chromosomal fibres (figs. 67a, 69a, 70a).

In great contrast to the single chromatid, haploid race of *H. diversa* there are many species of *Spirotrichonympha* where the conditions are the same as in this race of *H. diversa*, yet the sister chromatids pair intimately from one end to the other (fig. 73). The same situation occurs in many species of *Trichonympha*, for example in *T. grandis* with a haploid chromosome number of 24 (fig. 74). In fact, pairing of chromatids is the rule rather than the exception in hypermastigotes. In *Spirotrichonympha* and *Trichonympha* the reorganization of a new nuclear membrane from the old one does not begin until telophase instead of prophase as in *Holomastigotoides*; hence, the formation of the new nuclear membrane occurs too late in the first two genera to interfere in any way with pairing. Perhaps this is why pairing of sister chromatids is complete in these genera.

Of course, if one were to study only late prophases of *H. diversa* where pairing has already occurred, one might conclude—and with apparent justification—that the juxtaposed regions of the sister chromatids do not represent pairing; that such chromatids either have not divided completely or that they are in the process of separation following division. But one only has to look at earlier prophases—when the chromatids always lie entirely free and each has many minor coils with major coils just beginning to come in—to see this is clearly not the case (figs. 61, 62). These chromatids become apparently shorter as the gyres of the major coils become greater in diameter and fewer in number (fig. 63). Then pairing begins, and one can see plainly that the chromonema, which is now in the form of large major coils, of one sister chromatid firmly joins that of the other at certain common points (fig. 70). It might also be noted that at this time each gyre of the major coils has several gyres of minor coils incorporated in it (figs. 64, 71).

If one studies these paired sister chromatids, either in living or in fixed and stained material, from the standpoint of the matrix, one sees that the matrix of one chromatid is so intimately and firmly connected with that of its sister that the point where the two matrices meet can scarcely be detected (fig. 69).

2. HAPLOIDS WITH DOUBLE CHROMATIDS

a. Pairing of Sister Half Chromatids

When we turn our attention to the haploid race whose chromatids are double—each being composed of two half chromatids—we note pairing of the sister half chro-

---

which, following cytokinesis, new daughter nuclear membranes will be reconstituted. The rest of the old membrane will be discarded, mostly during metaphase and anaphase. The projection on the left is about to be discarded. Two terminal and one lateral nucleoli have been discarded.

66a. Slightly later stage. The short chromatids (light) have been cut off in separate portions or anlagen of the new nuclear membrane and lie far apart, while the long chromatids (dark), for the most part, lie in a common portion of a new nuclear membrane and are beginning to pair near their centromere ends. All nucleoli of long chromatids have been discarded. Note clear area around each centromere.

67a. A later stage than previous figure. Central spindle is longer and the two long (dark), partially paired chromatids are anchored to its posterior border, one at each end. The short chromatids have not made connection with the ends of the central spindle. They lie some distance apart in separate portions of the new nuclear membrane, while the long chromatids, for the most part, are in a common portion. All nucleoli have been discarded. The granular (stippled) portion of nuclear membrane is in the process of disintegration (portions are breaking away from it). Free astral rays omitted in this and all following figures except 68a.

68a. Later stage, yet there is no pairing of chromatids, since each chromatid was earlier cut off in a separate portion of the new nuclear membrane. Central spindle is almost completely unrolled and hence appears broader than in figs. 67a or 69a.

69a. This is a still later stage and this time the short chromatids are pairing in a common portion of a new nuclear membrane, while the long ones (dark) lie far apart in separate por-

tions. The granular region of the nuclear membrane is breaking up. All four chromatids are anchored to the ends of the central spindle; the short, paired ones being anchored to the anterior border (one at each end), and the long, unpaired ones to the posterior border. A free astral ray becomes a chromosomal fibre when it makes connection with a centromere.

70a. Slightly later stage than previous figure, long chromatids are partially paired, lie in centre, and are connected to the upper margin of the central spindle (one at each end), while the short (light) chromatids lie far apart and are connected to the lower margins of the central spindle (one at each end).

72a. Much like fig. 70a except a later stage in the disruption of the nuclear membrane.

73a. Late prophase of *Spirotrichonympha* sp. from *Rhinotermes* (= *Schedorhinotermes*) *intermedius intermedius* Brauer from New South Wales, Australia. Haploid (*n* = 2) single chromatid chromosomes. Centromeres are subterminal and one chromosome, which is considerably larger than the other, is drawn dark. Note that pairing of sister chromatids is continuous from one end to the other. Achromatic figure not drawn.

74a. Late prophase of *Trichonympha magna* Grassi from *Porotermes adamsoni* Froggatt, New South Wales, Australia. Chromatids are single and, for the most part, sisters are paired from one end to the other. This organism is usually a haploid. However, occasionally one sees what appears to be an example of aneuploidy. Note, at the lower left, what may be a tetrad; there seems to be a pairing of sister chromatids for a short distance and then of non-sisters. Small, early achromatic figure lies some distance anterior to nucleus.

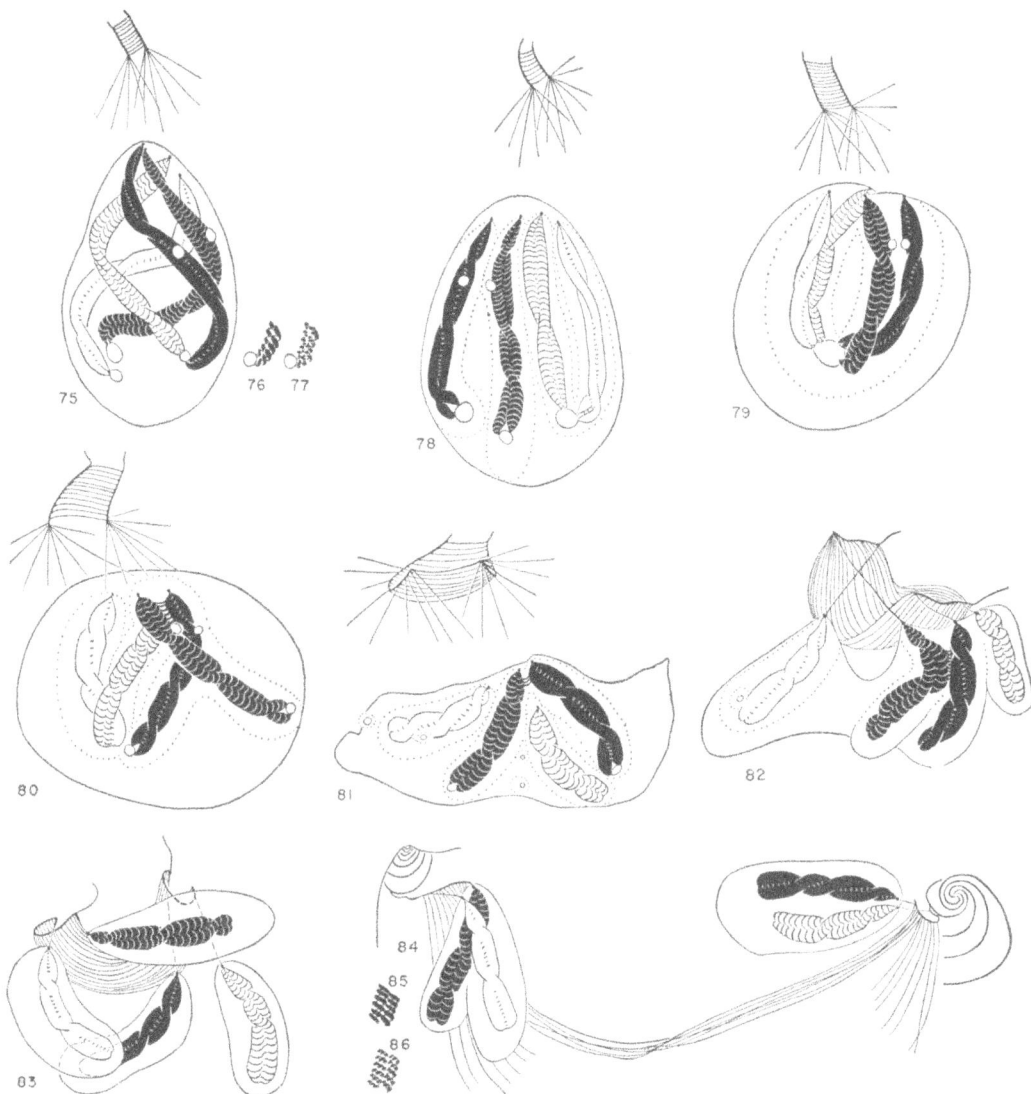

FIGS. 75–86. *Holomastigotoides diversa.* Haploid (*n* = 2). Double chromatids. × 1,200.

75. Fairly late prophase some time after duplication of chromatids (chromosomal duplication occurred in previous generation). Each of the two long (dark) and each of the two short (light) chromatids has a single centromere and is composed of two relationally coiled half chromatids which earlier were clearly separated but have now been brought together by pairing. After pairing begins, the relational coiling of half chromatids stops coming out.

76. Detail of portion of two half chromatids of fig. 75 to show their major coils.

77. Detail of a portion of two half chromatids of fig. 75 to show minor coils incorporated in majors.

78. Later prophase. Chromatids shorter and nearly straight

and each is being restricted to a small portion of the nuclear membrane which is destined to survive. Sister half chromatids still relationally coiled and paired, but there is no pairing of half-sister chromatids, and there will not be later, since they are already separated by their restrictive portions of the developing, new nuclear membrane except in the region of the fused nucleoli of the short (light) chromatids.

79. Later stage, but relational coiling of paired sister half chromatids has not come out and it will not do so during this cell generation, since they keep their pairing throughout (figs. 80–84).

80. The prophase relationally coiled sister half chromatids are still paired. Terminal as well as lateral nucleoli of long sister

matids and of the half-sister chromatids. We also note that pairing of the half chromatids begins earlier in the prophase than that of the chromatids and extends from one end to the other (figs. 75–80). In fact, it begins slightly earlier than the first figure of this group. There are always one or more gyres of relational coiling of sister half chromatids present when it begins and the pairing prevents these gyres from coming out. In fact, this relational coiling of sister half chromatids persists through nuclear division and cytokinesis and the pairing persists for some time after nuclear division (figs. 81–84). Each sister half chromatid has a common centromere (as do chromatids in meiosis I).

In the prophase when pairing of half chromatids begins each half chromatid has many individual major coils (fig. 76) and there are several gyres of minor coils incorporated in each gyre of the majors (fig. 77). Both major and minor coils persist throughout the various stages of nuclear division (figs. 85, 86). In fact, the minors are never lost and the majors do not begin to disappear until late telophase following cytokinesis (see Cleveland, 1949a, fig. 61b).

The new portions of the developing new nuclear membrane never cut in between the sister half chromatids. This is probably due to the fact that pairing of these half chromatids is always complete some time before the formation of a new nuclear membrane begins and thus prevents the membrane from cutting in between the paired half chromatids (figs. 75–82).

b. Pairing of Half-Sister Chromatids

If we turn to a consideration of the pairing of half-sister chromatids, we find there is no difference in the pairing of these chromatids which are double and in the pairing of sister chromatids which are single (figs. 63a–72a). Sometimes there is no pairing at all—probably owing to the fact that the portions of a new nuclear membrane cut in around each half-sister chromatid before pairing begins (fig. 78a). Then there may be a slight pairing of the long half-sister chromatids and no pairing of the short ones (figs. 81a, 82a), or there may be slight pairing in both long and short half-sister chromatids (fig. 80a). The main point to note in this race of H. diversa is the great difference between pairing of half-sister chromatids and that of sister half chromatids. Sister half chromatids invariably pair earlier and more intimately than half-sister chromatids.

Since we are dealing with a haploid organism (with two chromosomes), the pairing of the half-sister chromatids is always lost (pulled apart) as the two members of a pair move to opposite poles (figs. 82a–84a), while the pairing of sister half chromatids remains intact (figs. 82–84). In other words, there is segregation of half-sister chromatids, but not of sister half chromatids. The sister half chromatids have a common centromere and, as in chromatids in meiosis I, they do not separate unless crossing-over occurs.

3. DIPLOIDS WITH SINGLE CHROMATIDS

a. Pairing of Sister Chromatids

In the diploid, single chromatid race of H. diversa, it is interesting to note the chromatids begin to pair earlier in the prophase and the pairing is more complete than in a corresponding stage in the haploid, single chromatid race (compare fig. 63a with 87a). And there is always some pairing, in both short and long chromatids, in late prophases (89a). I cannot venture a guess as to why pairing begins earlier, and is more extensive, when four homologous chromatids are present than when there are only two (compare figs. 89a–92a with 65a–70a). In both instances, it is between chromatids. However, when only two homologous chromatids are present, as in haploids, pairing is always between sister chromatids; but, when four are present, it may be between sister chromatids or between non-sisters. There is, however, a difference in the pairing of sisters and non-sisters. Sisters may pair from one end to another, that is, pairing is complete, while non-sisters never pair completely (fig. 95a).

However, sometimes a prophase chromatid may pair with its sister in one region and with a non-sister in another region. This may occur at any point along a chromatid, either in long chromatids or short ones (fig. 90a). Such a pairing may be seen very plainly as the chromatids move poleward in anaphase (fig. 92a). As will be explained in a later section, this is the mechanism which produces an interlocked chiasma between homologous, non-sister chromatids. Crossing-over, as will also be explained in a later section, usually follows such a chiasma, but when sister chromatids pair from one end to the other, as in figure 95a, before pairing of non-sisters begins, a chiasma produced in this manner is usually not followed by

---

half chromatids are fused. Nucleoli of short (light) sister half chromatids have disappeared.

81. This prophase is much like the two previous ones (achromatic figure slightly larger and chromatids somewhat shorter). Sister half chromatids still paired and relationally coiled. Centromeres have made no connection yet with ends of central spindle.

82. This stage, which may be regarded as a metaphase, is slightly later than previous one. Achromatic figure is longer and is depressing the nuclear membrane, much of which will be discarded presently. Relationally coiled sister half chromatids paired from one end to the other. Free astral rays omitted.

83. Anaphase. Bent central spindle now lies in the centre of the four small, remaining portions of the original nuclear membrane. Sister half chromatids still paired and relationally coiled. Free astral rays omitted.

84. Later telophase with central spindle in process of pulling in two, yet sister half chromatids are relationally coiled and paired. Free astral rays omitted.

85. Detail of a portion of two half chromatids of fig. 84 to show their major coils.

86. Detail of a portion of two half chromatids of fig. 84 to show how the minor coils are incorporated in the majors.

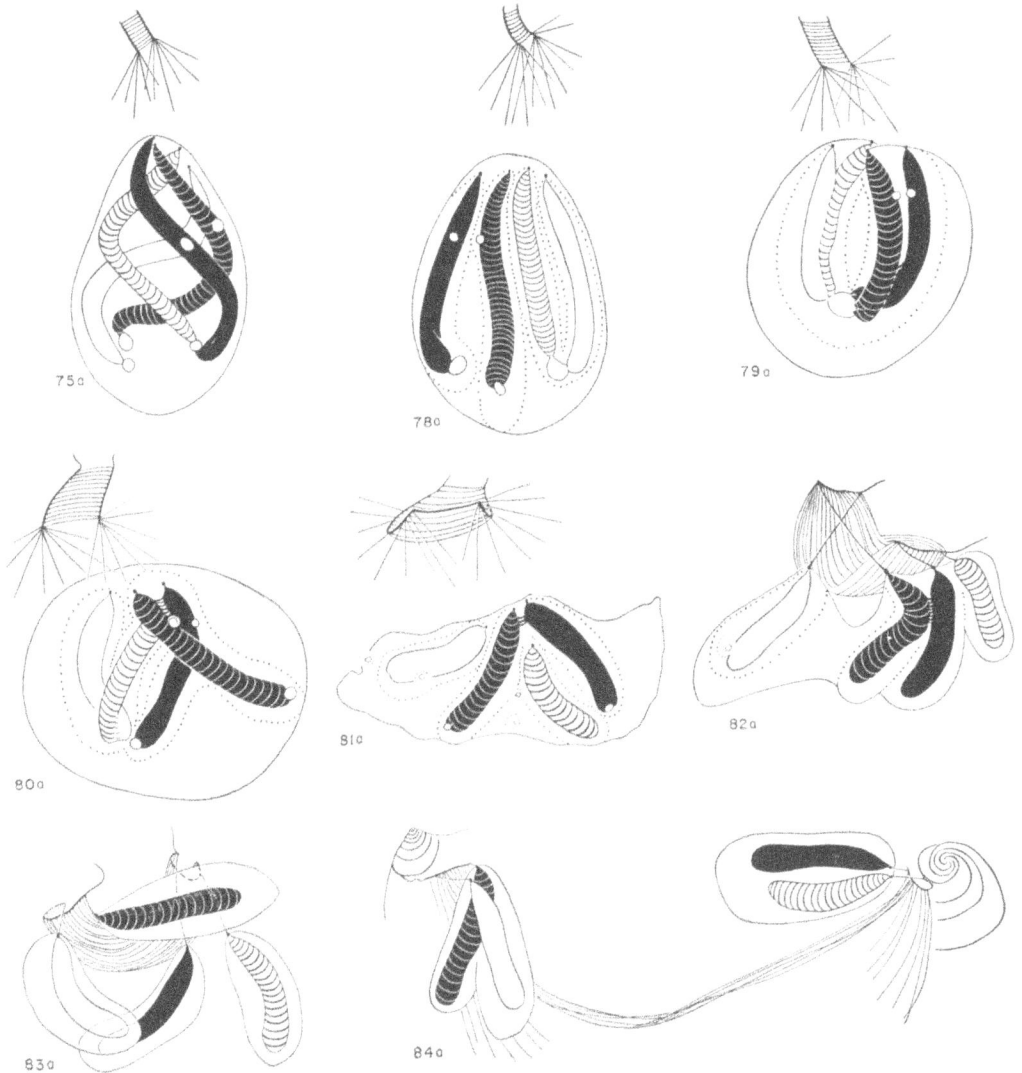

FIGS. 75a, 78a–84a.   Same as figs. 75, 78–84 except half chromatids are omitted in order to show more clearly the interrelationship
of half-sister chromatids when pairing and when not pairing, the disintegration of a large part of the parent nuclear membrane,
and the manner in which a new nuclear membrane is formed.   × 1,200.

75a. Late prophase. The long half-sister chromatids (dark) lie fairly close together, but are no longer relationally coiled about each other. The same is true of the short half-sister chromatids (light). No nucleoli have been lost.

78a. The chromatids are now nearly straight and are almost completely separated from one another by portions of the new nuclear membrane. Note long half-sisters on left and short ones on right. All nucleoli still present. No pairing.

79a. This is a later stage, from the standpoint of the chromatids, than the previous figure, yet the half-sisters are not in individual membranes. A common portion of the new nuclear

membrane is forming around the long (dark) half-sisters and another around the short (light) ones. There is a considerable amount of pairing between short half-sisters, but the long half-sisters have not begun to pair (one merely lies over the other, near their distal ends for a short distance).

80a. About same stage as previous figure. Note pairing in those regions of half-sister chromatids which are in a common portion of the new nuclear membrane and no pairing in the regions in separate membranes. Long half-sister chromatids paired at centromere ends; short ones, at distal ends.

81a. Several nucleoli have lost connection with chromatids

crossing-over. The interrelations, in mitosis and in meiosis, of chiasmata and crossing-over are considered in later sections.

When the prophase chromatids first begin to pair, each contains approximately twice as many gyres of major coils (fig. 87) as in metaphase or anaphase (figs. 93–98). As the chromatids become broader and apparently shorter, the gyres become greater in diameter. Each gyre of major coils contains several gyres of minors (figs. 88, 96). As the chromatids become broader and pairing becomes more intimate one can easily see that the major coiling systems of the paired chromatids are interconnected. Studies with dark phase contrast at this stage reveal many fine, hair-like extensions from each gyre. The chromonema is not so smooth as it frequently appears after fixation and staining. The matrix when viewed with phase contrast is also not smooth as chromatid pairing begins. It contains numerous bristles which extend in all directions. One can see the extensions of one chromatid, in the early stages of pairing, join those of another and thus, by means of interconnections formed in this manner, force the two chromatids to lie parallel, even though most of their surfaces are not in actual contact. Fixed and stained preparations of many organisms at this stage show parallel chromatids, but so far Barbulanympha and Holomastigotoides are the only ones in which the interconnections have been studied carefully.

These interconnections—which I plan to illustrate in detail in a later paper—are clearly the means of connecting the matrices of chromatids in the initial stage of the pairing process. In later stages of paired chromatids—both sisters and non-sisters—the matrices of homologous chromatids become so intimately and firmly connected that it is difficult to see the points where they first made contact or, indeed, to tell the matrix of one chromatid from that of another (figs. 93, 94).

b. Pairing of homologous non-sister chromatids

(= chromosome pairing)

Since the question of chromosome pairing is also considered later in connection with the diploid, double

chromatid race of H. diversa, it will be dealt with briefly here.

First, what is chromosome pairing? How, if at all, does it differ from chromatid pairing? More often in mitotic pairing sister chromatids pair from one end to the other or nearly so before the non-sisters begin to pair (fig. 89a). Strictly speaking, if the pairing of non-sisters were to bring the homologous chromosomes completely together, it would be regarded as complete chromosomal pairing, even though the paired chromosomes in reality would be brought together by the pairing of non-sister chromatids. But the pairing of non-sisters will not bring the chromosomes together unless the sister chromatids are also paired in one or more regions. Then, chromosome pairing, in final analysis, results from the pairing of both sister and non-sister chromatids. But what type of pairing is it when only non-sister chromatids pair in a certain region and sisters pair elsewhere? In the region, or regions, where only sisters are paired, the pairing cannot be regarded as chromosomal, for the chromosomes are not paired at all. This must be regarded as pairing of sister chromatids, since it is the same type of pairing as occurs in single chromatid, haploid organisms. But in those regions where only non-sister chromatids are paired, the pairing here is partial pairing of chromosomes, or non-sister pairing of chromatids, both of which are one and the same.

Complete pairing of chromosomes has never been observed in H. diversa. Figure 95a shows as complete pairing as I have ever seen. More often it is restricted to the distal ends (figs. 90a, 94a), but there is nearly always some pairing, either in the short or the long chromosomes, and occasionally it is fairly extensive in both short and long chromosomes (fig. 95a).

4. DIPLOIDS WITH DOUBLE CHROMATIDS

a. Pairing of Sister Half Chromatids

In organisms of this type there is pairing of sister half chromatids, of half-sister chromatids, and of non-sister chromatids or chromosomes. That of sister half chromatids begins slightly earlier than that of half-sister chromatids, and is usually completed shortly after the pairing of half-sister chromatids begins (fig. 99).

(only terminals of long ones remain attached). Pairing of long half-sister chromatids only in the region near the centromeres where they are in a common portion of the new nuclear membrane. Short half-sister chromatids lie far apart, each in its own membrane.

82a. Metaphase. All chromatids attached to central spindle (long ones at bottom; short ones at top). Pairing only in a short region of long half-sister chromatids where they are not separated by developing, common membranes. Remains of only one nucleolus left. Free astral rays omitted.

83a. Anaphase. All of parent or original membrane except a small portion around each chromatid has broken up and either has disintegrated in the cytoplasm or is in the process of doing so. Two of these regions (one surrounding a long and one a short chromatid) will fuse presently to form new daughter nuclear membranes. No pairing. Long (dark) chromatids con-

nected via their centromeres to upper margin of central spindle and short (light) ones to lower margin. Free astral rays omitted.

84a. Late telophase with central spindle in process of pulling in two. In the right hand daughter nucleus, the portions of the nuclear membrane which remained around the long (dark) and short (light) chromatids have fused to form a common, new nuclear membrane; in the left hand daughter nucleus, similar remaining portions of the original nuclear membrane are in the process of fusing to form a new one. The manner in which the centriole follows one of the four daughter flagellar bands (but leaves it before producing the achromatic figure) is shown in vertical view on right and lateral view on left. The new nuclear membranes will increase greatly in size shortly after cytokinesis. Free astral rays omitted.

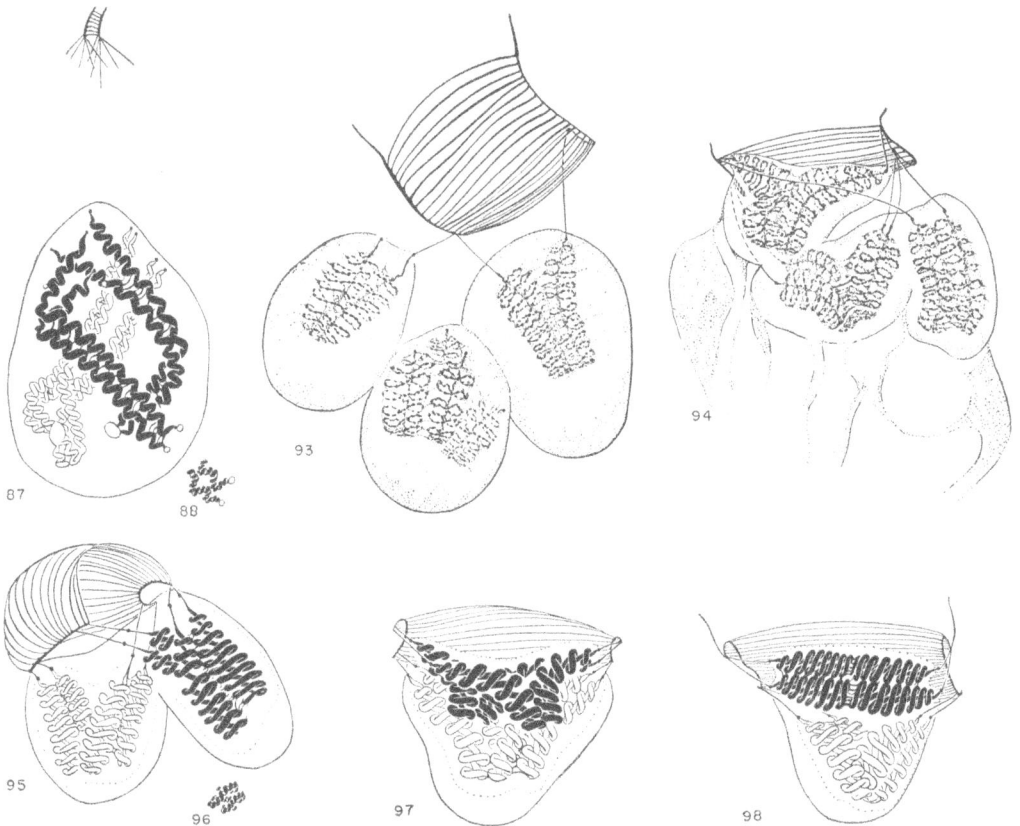

FIGS. 87, 88, 93–98. *Holomastigotoides diversa.* Diploid $(2n = 4)$. Single chromatids. Major coils, inserts of minor coils, and matrix in relation to coils and to pairing. × 1,200.

87. Prophase, after duplication of chromosomes and unwinding of chromatids (loss of relational coiling). Major coils shown in each chromatid. All four terminal nucleoli of the four short (light) chromatids fused; those of two sister, long (dark) chromatids fused. Centromeres of some sister chromatids lie fairly close together; others a good distance apart. Achromatic figure, which lies anterior to nucleus, has made no connection with centromeres.

88. Detail of a portion of the two long, sister chromatids of fig. 87 to show minor coils incorporated in majors.

93. Metaphase. Emphasis is placed on the matrix and the projections of the chromonemata. However, the big major coils, which loop back on themselves like figure 8's, may be seen. The extensions of the chromonema in this, and the next figure, do obscure to a considerable extent the very large number of minor coils that are incorporated in each major gyre. The four homologous short chromatids all lie in a common portion of the nuclear membrane in the centre, with two long, sister chromatids in a common nuclear membrane to the right and two to the left (see fig. 93a where chromatids are shown in technique). In the long sister chromatids, note how intimately the matrix of one sister is joined with that of the other, and how the sisters, in certain regions, appear to have a common chromonema (because of the intimate gyre to gyre pairing of corresponding portions of the sister chromatids). In the four

short chromatids, there is intimate pairing of sisters and some pairing of homologous non-sister chromatids. Not all 8 centromeres are connected via chromosomal fibres with the ends of the central spindle, those of the four short chromatids in centre in common portion of nuclear membrane and that of the long chromatid on upper left are unconnected. Note clear area around each centromere, and centrol rod extending from centromere to distal end of each chromatid. Free astral rays omitted.

94. This is an early anaphase drawn in the same manner as the previous figure. Two long, paired sister chromatids to right; two to the left, in which half the pairing has come out; and four short, homologous chromatids in centre. Each pair of long, sister chromatids lies in almost a separate portion of the nuclear membrane. The four short chromatids also lie together in almost a separate portion of the nuclear membrane. Note in long, sister chromatids to right intimate pairing of matrices and chromonemata from one end to the other (no poleward movement yet, even though each centromere is plainly connected with the central spindle; so, this pair of chromatids is metaphase). In other long, sister chromatids (upper left), poleward movement has separated the paired matrices and chromonemata for half the length of the chromatids (segregation). In the short, homologous chromatids (centre), the sisters, which lie on the right and on the left, are paired from end to end, while non-

Sometimes it is completed before half-sister chromatids begin to pair. Just as already noted in haploids with double chromatids, the pairing of sister half chromatids prevents further unwinding of these half chromatids. They have as many gyres of relational coiling during metaphase, anaphase, and telophase as at fairly late prophase when pairing is completed (figs. 99–117). Sometimes there are a few more gyres of relational coiling in the half chromatids of one long (or one short) chromatid than in the others, but on the whole the number does not vary much. The long half chromatids (dark) usually have one to two more gyres of relational coiling than the short ones (light). Their greater length easily differentiates them from the short ones from late prophase to mid-telophase, when their lateral nucleoli are absent. In late telophase and nearly all of prophase their lateral nucleoli clearly set them apart from the short ones.

The clear portions of the developing new nuclear membrane never cut in around the sister half chromatids and thus prevent them from pairing, because sister half chromatids are completely paired some time before these clear areas begin to develop. In fact, they usually do not come in before half-sister chromatid pairing is completed, or nearly so (figs. 104–105), and in many cases not before chromosome pairing has begun (figs. 106–109).

Sometimes, however, in addition to sister half chromatids being paired with each other from one end to the other, homologous, but non-sister, half chromatids also pair for a short distance (figs. 101, 107, 113). This is not the beginning of pairing of half-sister chromatids (figs. 99, 100, 104, 105). It may be regarded as the beginning of chromosomal pairing, since it is the means of bringing the two long homologous chromosomes together in this region.

As in the haploid race of *H. diversa* with double chromatids, the sister half chromatids of this diploid race always have a common centromere (for a discussion of this situation and its relation to meiosis, see Cleveland, 1949a: 16, 17).

From mid-prophase to almost the end of telophase each half chromatid has many gyres of major coils (figs. 102, 116) and each gyre of major coils has incorporated in it two or more gyres of minor coils (figs. 103, 117).

Perhaps I should point out that in the haploid and diploid races of *H. diversa* with double chromatids the gyres of major coils are smaller in diameter and more numerous in a given stage, say from late prophase through anaphase, than in a corresponding stage of the haploid and diploid races of this species with single chromatids (compare figs. 76 and 102 with 68 and 98).

b. Pairing of Half-sister Chromatids

In the diploid race of *H. diversa* with double chromatids, the pairing of chromatids begins much earlier in the prophase than in the haploid race with double chromatids (compare figs. 99a–106a with 80a–82a) or in the haploid race with single chromatids (figs. 66a–70a), but only slightly earlier than in the diploid race with single chromatids (fig. 87a). In the haploid race with single chromatids, homology is expressed twice; in the haploid race with double chromatids, four times; in the diploid race with single chromatids, four times; and in the diploid race with double chromatids, eight times. From the behavior under these four conditions, it is clear that the pairing of chromatids is influenced very little, if at all, by a chromatid increased expression of homology, but very greatly by a chromosomal increase in homology. In other words, chromatids pair much earlier and more completely in diploids than in haploids, irrespective of whether the chromatids are single or double.

In some instances, half-sister chromatids pair with each other from one end to the other in fairly late prophase (figs. 99a, 100a), but sometimes—as already noted in the description of pairing of half chromatids—there are regions where the half-sisters do not pair with each other; instead, the homologous non-sister chromatids pair with each other (via their non-sister half chromatids), and in so doing produce an interlocked chiasma (figs. 101a, 107a). In the pairing of half-sister chromatids, the centromere ends usually pair last. In fact, they often remain unpaired for some time after these chromatids are intimately paired elsewhere (figs. 101a–106a). Occasionally, the distal ends of half-sister chromatids behave in the same manner (fig. 105a). The unpaired ends—either centromere or distal—of half-sister chromatids sometimes are the means of initiating chromosome pairing (fig. 101a) which, of course, is the same as pairing of homologous, non-sister chromatids. At other times, the regions where half-sister chromatids are not paired, but homologous non-sister chromatids are paired, are the only pairing there is of chromosomes (fig. 113a). However, more often, perhaps, the half-

---

sisters are paired only for a short distance at their distal ends. The matrix and chromonema connections of non-sisters are being broken by poleward movement, while those of sisters remain intact (no segregation). Nuclear membrane is breaking up. Free astral rays omitted.

95. Later anaphase. Emphasis is on major coils. Matrices not shown. Note how the large major coils of one chromatid are paired with those of its sister or those of its homologous non-sister, the joining of one chromatid to another always being at precisely the same region on each. Free astral rays omitted.

96. Detail of a portion (two gyres) of a long (dark) chromatid of fig. 95 to show the large number of minor coils incorporated in each gyre of major coils.

97. Anaphase. Note large major coils in each chromatid and the manner in which the gyres of sisters are interconnected (paired). Free astral rays omitted.

98. Late anaphase. Note how intimately the major gyres of the long (dark), sister chromatids are paired from their centromeres almost to their distal ends. At distal ends, the terminal gyres of non-sisters are joined. Free astral rays omitted.

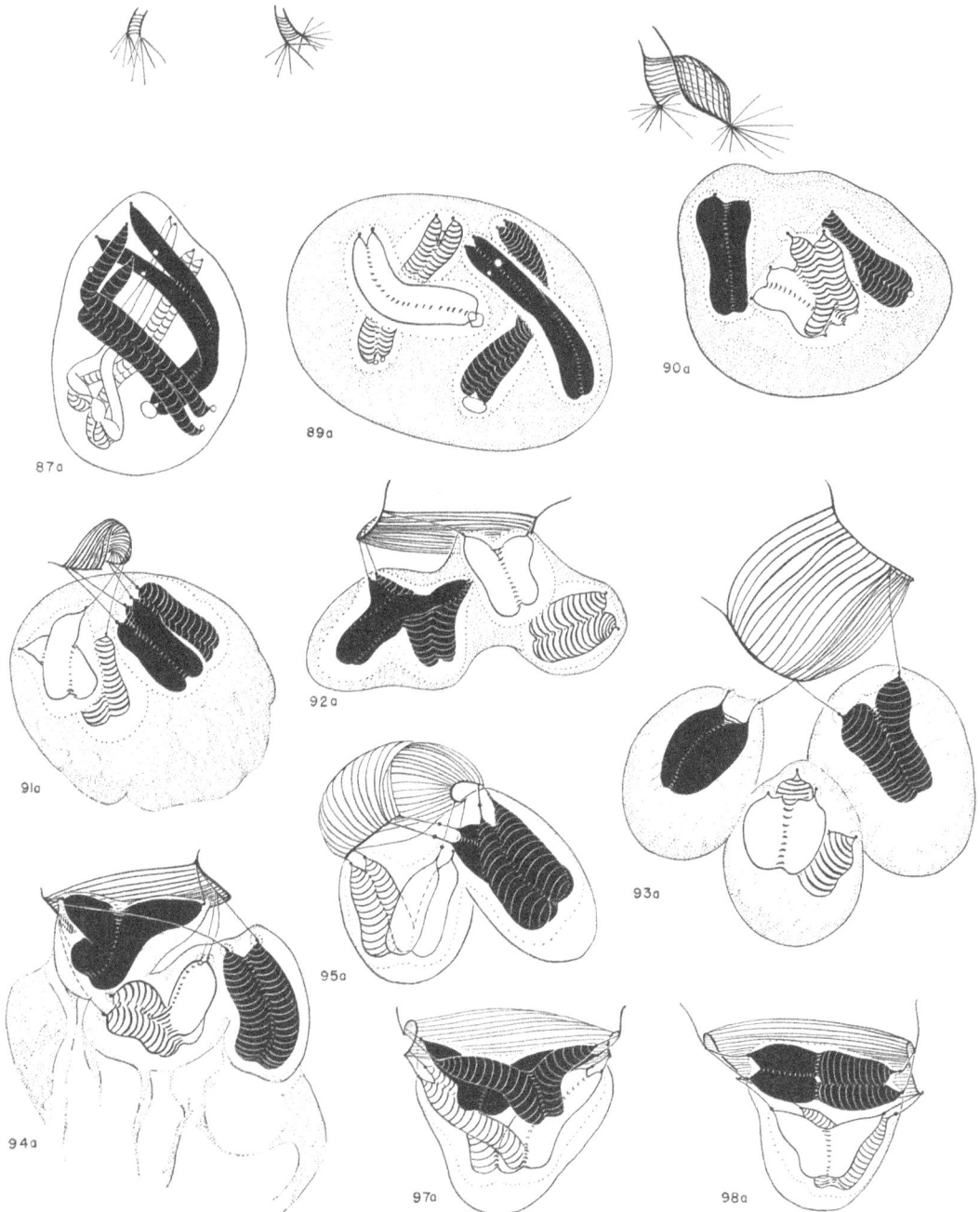

FIGS. 87a, 89a–95a, 97a, 98a. *Holomastigotoides diversa.* Diploid (2n = 4) single chromatids. Major coils have been omitted in order to show more clearly the relation of the sister chromatids and of the homologous chromosomes to each other. × 1,200.

87a. Note pairing in certain regions of the sister chromatids in both of the long (dark) and in both of the short (light) chromosomes. All terminal nucleoli of short chromosomes fused; those of one pair of long chromatids fused, but not of the other (lower right). No pairing of chromosomes yet.

89a. Later prophase. The sister chromatids, in each of the four chromosomes, are now paired from one end to the other, but there is no pairing of homologous, non-sister chromatids (= chromosome pairing, for example, one long (dark) chromosome with another; or one short (light) chromosome with another). However, the short chromosomes lie near each other on the left, and the long ones on the right. Each chromosome lies in its own portion of the new nuclear membrane. The long chromosome on the right has lost its terminal nucleoli. Achromatic figure is still small and lies a considerable distance from the nucleus.

90a. Late prophase or early metaphase. Each long (dark) chromosome lies in its own portion of the new nuclear membrane, while both short chromosomes lie in a common portion. There is no pairing of long chromosomes (non-sister chromatids); in fact, they lie far apart, but sister chromatids of each chromosome are intimately paired. In the short (light) chromosomes, sister chromatids of one chromosome (all light) are paired from one end to the other; but in the other short chromosome, sister chromatids are paired from the centromere ends about two-thirds the way toward distal ends. From this point on to the distal ends, there is no pairing of sister chromatids; instead, non-sisters are paired (= chromosome pairing). In the mid-region of the short chromatids, there is pairing of both sister and non-sister chromatids for almost half their length, but the pairing is confined to the two chromatids in the centre. One of these chromatids is paired with three other chromatids: its sister, which lies to the right of it, in the centre with a non-sister, and near the distal end with another non-sister. Achromatic figure has increased in size but has not made contact with the centromeres.

91a. Slightly later than previous stage. Some of the free astral rays have become chromosomal fibres by making connection with the centromeres. The four long (dark) chromatids lie in a common portion of the new nuclear membrane, while in the short chromatids (light) only the sisters lie in common portions. Sister short chromatids are paired for most of their length. In each case, only one sister is connected to the end of the central spindle. In the long chromatids, there is pairing of both sisters and non-sisters, but that of sisters is more extensive. Non-sisters are anchored to the same end of the central spindle which means there will be segregation of sister chromatids at anaphase. Free astral rays omitted.

92a. Slightly later stage of development than the previous figure and the pairing of chromatids is quite different. In the long chromatids, all of which lie in a common portion of the new nuclear membrane, the distal two-thirds of sisters are paired, but at the centromere ends non-sisters are paired for one-third their length. Poleward movement is underway in all four chromatids. Non-sisters are going to the same pole (end of central spindle), and, since the distal two-thirds of sister chromatid pairing is stronger than the anterior one-third of non-sister pairing, it is obvious that the two inner non-sisters will break at the point where they are interlocked. But before the break is complete, a union between non-sisters begins (see figs. 125, 132) when the break is in progress, so there are never any entirely free ends of chromatids. In a slightly later stage than the one shown, the distal two-thirds of sister chromatids and the anterior third of non-sisters would be going to the same pole. In other words, the anterior third of the two non-sisters will have crossed over, while the distal two-thirds of these same

chromatids did not. This is segregation in one region and non-segregation in another.

The two short (light) chromosomes lie in separate portions of the new nuclear membrane. In the upper chromosome (all light), the sister chromatids are going to opposite ends of the central spindle, and, as a result of their poleward movement, their pairing is being resolved. In the lower chromosome to the right, sister chromatids are paired from one end to the other, and these chromatids have not made connection with the end of the central spindle. However, when they do, they will have to connect with opposite ends of the central spindle, owing to the fact that the chromatids of the other homologous chromosome have already been connected in this manner. Thus, the chromatids of one homologue set the behavioral pattern for those of the other homologue. Free astral rays omitted.

93a. This nucleus is in about the same stage of development as the two previous ones except for the behavior of the nuclear membrane, which has broken up into three almost complete portions. Each portion, however, will become much smaller later. The sister chromatids of each of the two long (dark) chromosomes—which lie some distance apart—are paired. The pairing is being resolved by their movement to opposite poles (segregation). In the four short (light) chromatids, which lie close together, there is pairing of sisters and non-sisters (= chromosome pairing). Free astral rays omitted.

94a. Anaphase. One pair of the long (dark) chromatids is going to opposite poles and the other, which is connected to opposite poles, will do likewise presently. No pairing of long chromosomes. In the four short chromatids, all of which lie in a common portion of the nuclear membrane, poleward movement is in progress and the pairing of chromosomes (= non-sister chromatid pairing) is being pulled apart by this movement. The sister chromatids are going to the same pole without losing any of their pairing, which is from end to end. Thus, there is segregation in the sister long chromatids, but not in the sister short ones. Free astral rays omitted.

95a. In this anaphase there is chromosomal pairing in the two short (light) as well as in the two long (dark) chromosomes. At present, it is more extensive in the long chromosomes. One can see, however, from the manner in which the chromatids of each chromosome are anchored to the poles and the way their centromere ends are directed, that there will be no segregation. Sister chromatids, in every instance, will go to the poles while they maintain their pairing from one end to the other. If the chromosomal pairing—which in reality is nothing more than pairing of non-sister chromatids—were as extensive as that of sister chromatids, then one could not determine whether segregation occurs or not, because sister and non-sister chromatids could not be differentiated from each other. Fortunately this is not the case. Free astral rays omitted.

97a. A later anaphase. In each of the two long (dark) chromosomes, sister chromatids are going to opposite ends of the central spindle (segregation), but in each of the short (light) ones they are going to the same end of the central spindle while paired from end to end (no segregation). Free astral rays omitted.

98a. In this anaphase, which is somewhat later than the previous one, there is no segregation in the chromatids of the two long (dark) chromosomes, while there is segregation in the chromatids of the two short (light) chromosomes—just the reverse of the previous figure. In each long chromosome, sister chromatids are paired from centromere to distal ends. The pairing, which holds the distal ends of the chromatids together, is between non-sister chromatids. This is chromosomal pairing, which earlier may have been more extensive. There is no chromosomal pairing in the short (light) chromosomes (the two entirely light ones merely lie partly over the others). Free astral rays omitted.

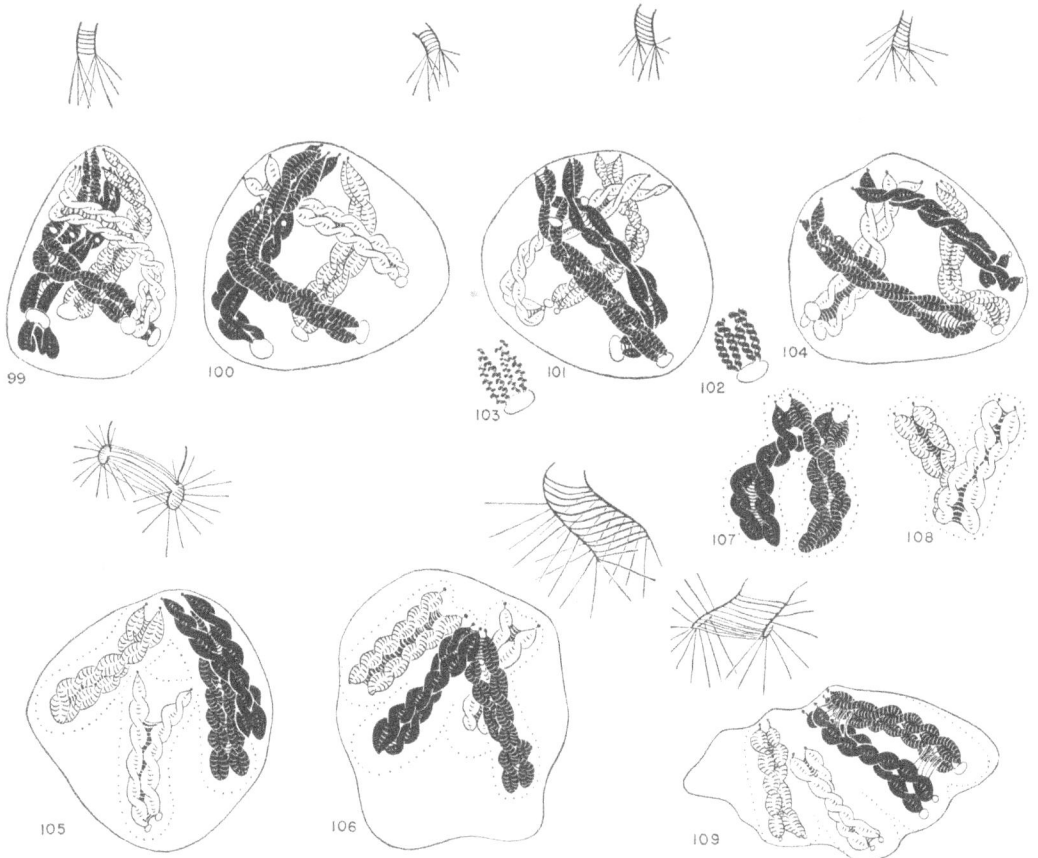

FIGS. 99–109. *Holomastigotoides diversa.* Diploid ($2n = 4$). Double chromatids. In most drawings proximal ends of centrioles are omitted. × 1,250.

99. Prophase, some time after duplication of chromatids since most of the relational coiling of half chromatids has come out (unwinding) and no more will come out because pairing of sister half chromatids (which is now continuous, in each chromatid, from one end to the other) will prevent further unwinding. Pairing is nearly complete (from one end to the other) in most of the half-sister chromatids, of which there are 8, four homologous, long (dark) ones and four homologous, short (light) ones, but pairing of homologous chromosomes has not begun. Terminal nucleoli of homologous chromatids are fused.

100. Slightly later prophase. Sister half chromatids, as in previous figure, are paired for their entire length and are relationally coiled.

101. About same stage of development as previous prophase, but pairing in the four long (dark) chromatids has occurred in a different manner. From their lateral nucleoli to their distal ends half-sister chromatids are paired, but from the lateral nucleoli to the centromere ends this is not the case. In this region, homologous, non half-sister chromatids are paired, thus producing an interlocked chiasma. See fig. 101a where the same type of pairing is shown with half chromatids omitted. Sister half chromatids are paired as in two previous figures.

102. Detail of the distal ends of four of the long (dark) half

chromatids to show their individual major coils.

103. Same as previous figure except both minor and major coils are drawn to show how minors are incorporated in majors.

104. Prophase in about same stage of development as figs. 100, 101. Pairing of half chromatids is complete, and that of chromatids except for small regions, mostly at centromere ends, is also complete. Note that the two long (dark) chromosomes lie a considerable distance apart, and the short ones, for the most part, are also far apart—hence, no chromosomal pairing.

105. This is a later prophase. Achromatic figure has begun to grow and chromatids are shorter and straighter. New, restrictive portions of the nuclear membrane have cut in around each short (light) chromosome and have thus prevented pairing of these chromosomes. The long (dark) ones lie in a common, restrictive portion of the new nuclear membrane and are pairing, but not so as to produce interlocked chiasmata.

106. Slightly later prophase. The two short (light) chromosomes lie far apart in separate, restrictive new portions of the nuclear membrane and the two long ones, for the most part, are also in the same condition. However, near their centromere ends they are in a common membrane and in this region they are pairing, but again not so as to produce interlocked chiasmata. Sister half chromatids paired and relationally coiled.

sister chromatids are paired from end to end before pairing of homologous, non-sister chromatids begins (figs. 106a, 109a). As pairing of homologous, non-sister chromatids continues, the four chromatids are brought together like four sausages, so that any way they are viewed two are on top and two on the bottom (fig. 112a). It is perfectly clear that such a configuration can be brought about only by a pairing of chromatids, and in this order: first, half sisters pair from end to end; second, non-sisters pair for about three-fourths of their length. More often the pairing of non-sisters is for a shorter distance, never for a greater one.

Now, how does the pairing of chromatids that produces an interlocked chiasma differ from that illustrated in figure 112a which has nothing even remotely suggesting such a chiasma? The same four chromatids take part in each process, but they cannot produce an interlocked chiasma if the half-sisters pair first from one end to the other. In order to produce such a chiasma, the half-sisters not only must pair in one region of the homologous chromosomes and the non-sisters in another (figs. 101a, 107a), but, in one region, the non-sisters must pair before the half-sisters do; while in other regions, namely, those where the half-sisters are paired, this pairing may occur before that of the non-sisters, at the same time as the non-sisters, or later. The whole point is, an interlocking between the chromatids of homologous chromosomes is produced only when non-sister chromatids pair before half-sisters do in a given region. Also, it perhaps should be noted at this point that the pairing of half-sister chromatids at other regions is just as necessary in the production of the interlocking as the pairing of non-sisters. In brief, an interlocked chiasma is produced by half-sisters pairing in one region and non-sisters in another.

Chiasmata in which chromatids are not interlocked may be responsible for crossing-over, especially in meiosis. The two types of chiasmata and their relation to crossing-over are considered in a later section.

As the chromatids move in opposite directions in the anaphase, one type of chromatid pairing is resolved (pulled apart). This may be either that of half-sisters (fig. 111a) or of non-sisters (figs. 112a, 113a), depending, in general, on whether there is chromosomal pairing or not and, in particular, as we shall see in a later section, on the manner in which the chromosomal pairing occurs. The pairing that is not removed by poleward movement remains for a short time after nuclear division (figs. 114a, 115a), but disappears before cytokinesis.

c. Pairing of Homologous Non-sister Chromatids (= Chromosome Pairing)

As explained already, the pairing of homologous chromosomes is brought about by—in fact is the same thing as—the pairing of non-sister chromatids (or, as in figs. 101, 107, of non-sister half chromatids).

There is usually some pairing of chromosomes, but it is never complete—from end to end—and for this reason one can tell the pairing of non-sister chromatids from that of half-sisters, since any chromatid pairing that is complete is of half-sisters. In other words, one can differentiate the pairing of chromosomes (or non-sister chromatids) from that of half-sister chromatids by the completeness of the pairing. Occasionally, there is no chromosome pairing at all (fig. 111a); sometimes there is pairing of the long chromosomes and none of the short ones; sometimes it is the other way around; and sometimes there is pairing in both long and short chromosomes. And the pairing may vary from a slight joining of the distal ends to intimate pairing for two-thirds the length of the chromosomes.

MEIOTIC PAIRING

In so far as our observations will permit us to generalize, it seems meiotic chromosomes usually begin to pair earlier in the prophase than mitotic ones. Of course, most mitotic chromosomes do not pair. I am thinking only of those that do. And I am also leaving out of the present consideration those organisms that complete meiosis in a single nuclear division, owing to the fact that centromere and chromosome duplications are inhibited simultaneously.

So far as I know, there is only one organism where the whole life cycle of the meiotic chromatids may be seen clearly in all stages from the time of chromosomal duplication, through unwinding of the relationally coiled chromatids, the formation of major coils, the various steps in pairing of sister and non-sister chromatids, crossing-over, segregation, poleward movement, nuclear division, cytokinesis, the loss of the major coils in very late telophase, the resting stage at the end of meiosis I, and in meiosis II the formation of major coils, pairing of chromosomes (no chromatids in II), segregation of chromosomes, poleward movement, nuclear division, cytokinesis, loss of the major coils, and to the resting stage at the end of meiosis II. These chromatids are so very plain that it is not only possible to see them easily in living cells; they may be photographed (see last paper of this series, photos. 169–173, 178–179, 190–193).

---

107. In this prophase tetrad there is chromosome pairing only in the region near the centromeres where both chromosomes are in a common portion of the developing, new nuclear membrane. And the chromosomal pairing is between homologous, non half-sister chromatids, thus producing an interlocked chiasma.

108. In this prophase tetrad pairing of sister half chromatids and of half-sister chromatids is continuous from end to end.

There is chromosomal pairing, but only in a region near their distal ends where both chromosomes are in a common portion of the developing, new nuclear membrane.

109. In this prophase nucleus, pairing of half chromatids and of half-sister chromatids is continuous, and chromosomal pairing is beginning between the two long (dark) chromosomes, but it cannot result in the production of interlocked chiasmata.

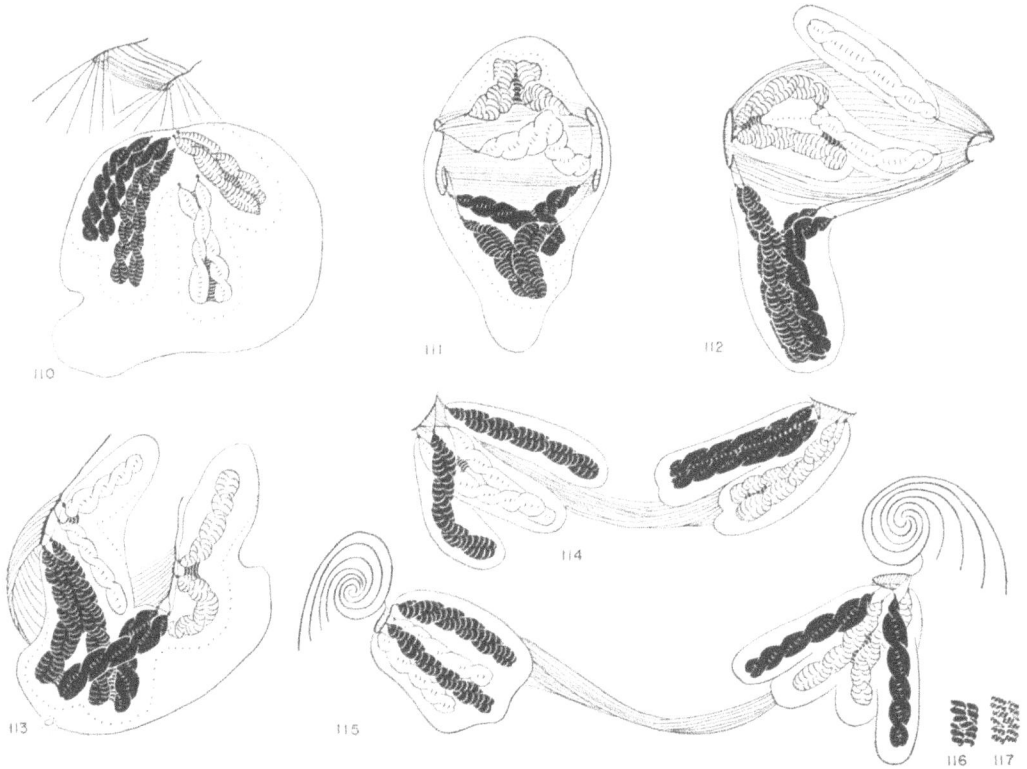

FIGS. 110–117. *Holomastigotoides diversa.* Diploid $(2n = 4)$. Double chromatids. In some drawings proximal ends of centrioles are omitted. Free astral rays omitted in all figures except 110. × 1,200.

110. In this prophase, which is as late as that of fig. 109, there is no pairing of chromosomes; sister chromatids and half-sister chromatids only are paired. Yet the long (dark) chromosomes lie close together in a common portion of the new nuclear membrane.

111. In this anaphase there is no pairing of chromosomes (= non-sister chromatid pairing). The pairing of the half-sister chromatids is being resolved by poleward movement. It is about half out (earlier it probably extended from end to end as in figs. 109, 110). The pairing of sister half chromatids is not being resolved, cannot be since these chromatids have a common centromere. There is segregation of chromatids; two homologous, non-sister long (dark) ones and two homologous, non-sister short (light) ones are going to each pole (centriole).

112. This anaphase, which, on the whole, is slightly later than the previous one, presents a very different situation. There is almost complete pairing of the two long (dark) chromosomes. These chromosomes are in a common portion of the new nuclear membrane which is free from that surrounding two of the short (light) chromatids (upper right) and that surrounding one short (light) chromosome (upper left). The short chromatids have progressed more rapidly in poleward movement than the long ones. They are in late anaphase or early telophase while the long ones are in early anaphase (a good portion of the nuclear membrane has been discarded). There was probably never more than a slight pairing of the short chromosomes. In the tetrad formed by the long (dark) chromosomes, it is clear that

half-sister chromatids (which are paired from one end to the other) are going to the same poles (centrioles). The same has already taken place in the half-sister, short chromatids. Hence, there is no segregation. In the tetrad formed by the pairing of the two long (dark) chromosomes, there are three types of pairing, that of sister half chromatids, of half-sister chromatids, and of non-sister chromatids.

113. This anaphase, in most ways, is in a slightly later stage of development than the previous one. Although some of the nuclear membrane has been discarded, and more will be discarded presently, the short (light) chromatids are at the poles and very little pairing of half-sisters remains. The long (dark) chromatids and chromosomes, which earlier probably formed a tetrad in which chromosomal pairing was somewhat more extensive, are moving poleward, thus breaking up the tetrad formation by removing chromosomal pairing while permitting most of the pairing of half-sister chromatids to remain. It should be noted, however, that the pairing of homologous, non half-sister chromatids at the distal ends is being resolved while that of the half-sisters at the centromere ends remains intact. Thus, the interlocking produced earlier in chromosome pairing by non-half-sister chromatids pairing in one region (at distal ends) and half-sisters in another region (at centromere ends) is being resolved without breakage and crossing-over occurring. The longer, stronger pairing union or bond is winning out over the shorter, weaker one by a process which has been called terminalization. There is no segregation.

114. Telophase after nuclear division. Sister half chromatids

The organism in which chromatids are so very plain in all stages is *Barbulanympha*. I have studied meiosis in the four species of this genus for many years and am preparing a detailed account of the process. However, since it is much too long to include as a part of the present paper, only a brief account of prophase pairing will be given.

As already shown (Cleveland 1949a, fig. 55) the chromatids of *Barbulanympha*, in mitosis and in meiosis, are single. Thus at the time of chromosome duplication in prophase I, each chromosome produces two homologous sister chromatids which have many gyres of tight relational coiling (presenting an appearance much like that of fig. 12b, Cleveland, 1949a). The sister chromatids immediately begin to unwind and the gyres of relational coiling become farther apart (photos. 149, 169). Finally, most of the relational coiling of the sister chromatids comes out; only a gyre or so, as a rule, remains as the homologous chromosomes move close together and begin to pair (photos. 170–172). In some instances, after pairing is completed, there are still one or two gyres of relational coiling in the chromatids of each chromosome, while in others there is less than a gyre (photo. 173).

Observations on both living and fixed materials show that these meiotic prophase chromatids of *Barbulanympha* pair in the same manner as the mitotic chromatids of the diploid single chromatid race of *Holomastigotoides diversa* (figs. 89a–92a). Sometimes, the sister chromatids of each homologous chromosome pair from one end to the other before the non-sisters begin to pair (fig. 135). In other words, chromosome pairing does not begin until sister chromatid pairing is complete (fig. 136). In such cases, non-sister pairing produces no interlocking of two chromatids—one from each chromosome. However, it does produce what has been called a chiasma (fig. 137). This makes it necessary, then, to recognize two types of chiasmata, which differ in their manner of formation as follows: (1) a chromatid of one homologous chromosome pairs with that from the other chromosome in a certain region before the pairing of sister chromatids occurs in this region, and thus a chromatid interlocking is produced (fig. 139); (2) the sister chromatids of both homologous chromosomes pair with each other from one end to the

other before pairing of non-sister chromatids begins and thus makes a chromatid interlocking impossible (figs. 135–137), but nevertheless produces a chiasma.

SEGREGATION [2]

Studies in recent years on centromeres in mitosis and meiosis have contributed much toward a better understanding of the differences in segregation of chromatids and chromosomes in the two types of nuclear division.

In haploid nuclei whose chromatids are single, there is no choice in segregation, for sister chromatids, irrespective of whether they pair or not, must go to opposite poles during anaphase and telophase (figs. 63a–72a). They have no choice, because when the centromere of one chromatid becomes connected with a pole via a chromosomal fibre, there is no longer any attraction between this pole and the centromere of the other sister chromatids; this centromere and chromatid must go to the opposite pole. If there is only one pole or centriole present, as occurs fairly often in some genera of flagellates, both sister chromatids never go to it. This has been observed over and over again, in both living and fixed cells.

There is also no choice in segregation of chromatids in haploid nuclei with double chromatids (each chromatid is composed of two sister half chromatids which share a common centromere). Segregation of these chromatids—which must be regarded as half-sisters—is the same as in those that are single, for the centromeres, which, in this case, are the sole controlling mechanism in segregation, are haploid (figs. 78–84). As a matter of fact, in organisms of this type one has to decide whether centromeres or chromatids are the determining factor in haploidy and diploidy. Such or-

[2] The term segregation throughout this paper is used in a slightly broader sense than in genetics. It refers to the movement to opposite poles during mitosis of sister chromatids, of half-sister chromatids, and of homologous chromosomes, as well as to the movement to opposite poles of homologous chromosomes during meiosis. Segregation when used in this sense renders unnecessary the use of the confusing and cumbrous term "equational division." The elimination of this term from cytology makes possible the further elimination of the equally confusing and ambiguous term "reductional division," as explained in the section devoted to discussion. Such changes and clarification in terminology also make crossing-over and partial segregation synonymous, the latter being the better term.

are still relationally coiled and paired from one end to another. Fusion is occurring in the small, independent portions of the original nuclear membrane which were not discarded and which now surround either a chromosome or a chromatid. In this manner a new nuclear membrane is reconstructed from the surviving portions of the old one.

115. A considerably later telophase. The central spindle, which extends from the anterior end of one daughter nucleus to that of the other, will soon disappear. Fusion of remnants of old nuclear membrane is complete in left hand daughter nucleus, and nearly so in that on right. However, these daughter nuclei will continue to grow and increase in size until cytokinesis and sometimes longer (compare figs. 110 and 115). Except for the short (light) chromatids in right hand daughter

nucleus, all pairing of chromatids has been lost. But pairing of sister half chromatids remains intact from end to end of each chromatid; it is lost shortly before cytokinesis. One cannot tell, in so late a stage, whether there was segregation of the long (dark) chromatids or not. The drawing has been made as if no segregation occurred, since this is more often the case. Note relationship of proximal end of a centriole to one of the four flagellar bands in each (widely separated) group of daughter bands.

116. Detail of the distal end of two long (dark) sister half chromatids to show their individual major coils still intact and much the same in appearance as in prophase (see fig. 102).

117. Same as previous figure except both minor and major coils are drawn to show how minors are incorporated in majors.

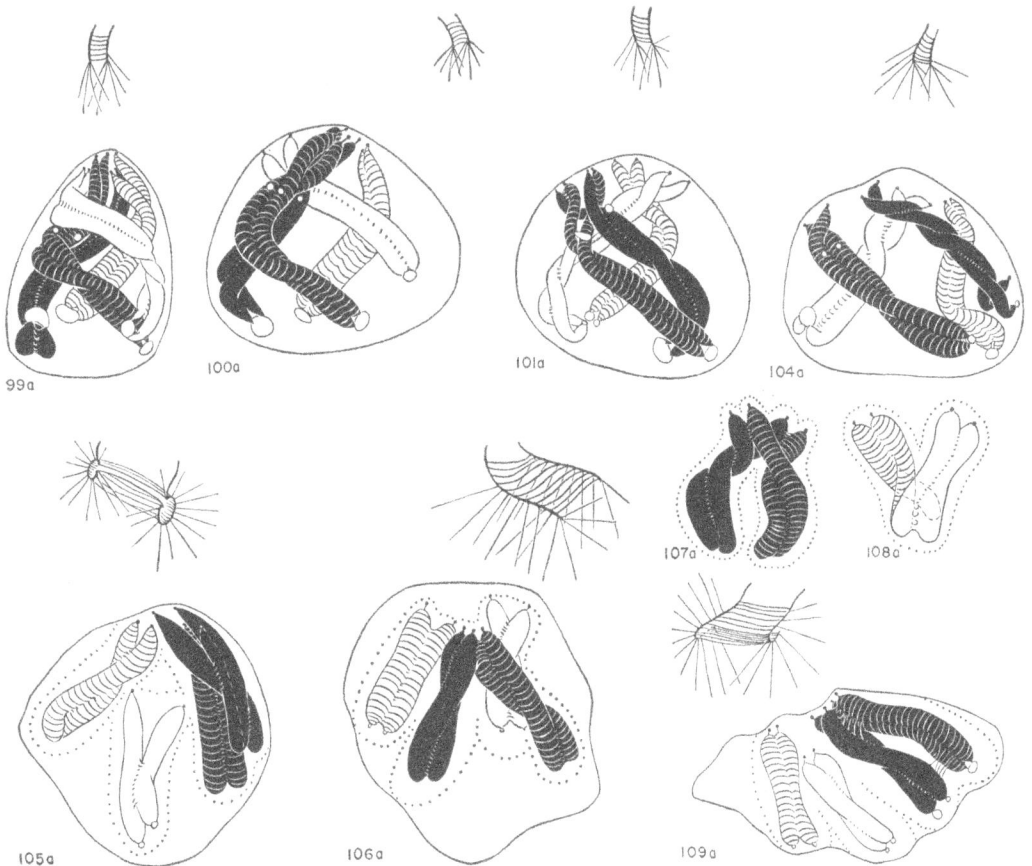

FIGS. 99a–101a, 104a–109a. Same as figs. 99–101, 104–109 except half chromatids are omitted in order to show more clearly the relation of chromatids to chromosomes in pairing, tetrad formation, chiasmata, and segregation. Even though half chromatids are not shown in these figures, the chromatids drawn must be referred to as half-sisters rather than sisters, if confusion is to be avoided. Proximal ends of centrioles omitted in most of the drawings. × 1,250.

99a. There is still about one gyre of relational coiling in the chromatids of each chromosome and, except for a short region at their centromere ends, and in some cases a similar region at their distal ends, the half-sister chromatids are paired. Thus, there is no difficulty in distinguishing, for example, one long (dark) chromosome from the other. The same is true of the short (light) chromosomes.

100a. There is no relational coiling of the half-sister chromatids of the short (light) chromosomes and very little in the long (dark) ones. The half-sister chromatids are now paired almost from one end to the other. Even though the centromeres and chromatids of the long chromosomes lie closer together than in the short chromosomes, there is no chromosome pairing (= pairing of non-sister half chromatids).

101a. Note pairing in the two long (dark) chromosomes of non-half-sister chromatids in centromere regions and of half-sisters elsewhere, thus produing an interlocked chiasma.

104a. The two long (dark) chromosomes, like the two short (light) ones, lie a considerable distance apart.

105a. Short (light) chromosomes far apart; long (dark) ones close together. Half-chromatids of each chromosome are paired; in some cases for most of their length; in others for only half.

106a. Same as previous figure and long (dark) chromosomes are pairing, but, so far, only two chromatids are taking part in the process. Half-sister chromatids are paired for most of their length.

107a. This is a later stage of a tetrad similar to the earlier one shown by fig. 101a.

108a. There is pairing of the short chromosomes at their distal ends, but no interlocked chiasma is formed because the half-sister chromatids of the other chromosome (the one on the right) are paired from one end to the other.

109a. Here in the two long (dark) chromosomes there is chromosomal pairing at both centromere and distal ends, but there are no interlocked chiasmata since pairing of the half-sister chromatids, in each chromosome, is from end to end. The two short (light) chromosomes lie some distance apart and are in nearly independent portions of the developing new nuclear membrane.

ganisms are clearly diploids in one sense and haploids in another. From the standpoint of terminology, and for other reasons too, it seems best to determine haploidy and diploidy on a basis of centromeres. However, two-division meioses present a problem, because at the end of the first division the two nuclei from the standpoint of the centromeres are haploid, since centromeres are not duplicated, while from the standpoint of the chromosomes, which are duplicated, they are diploid. One cannot get around this awkward terminological

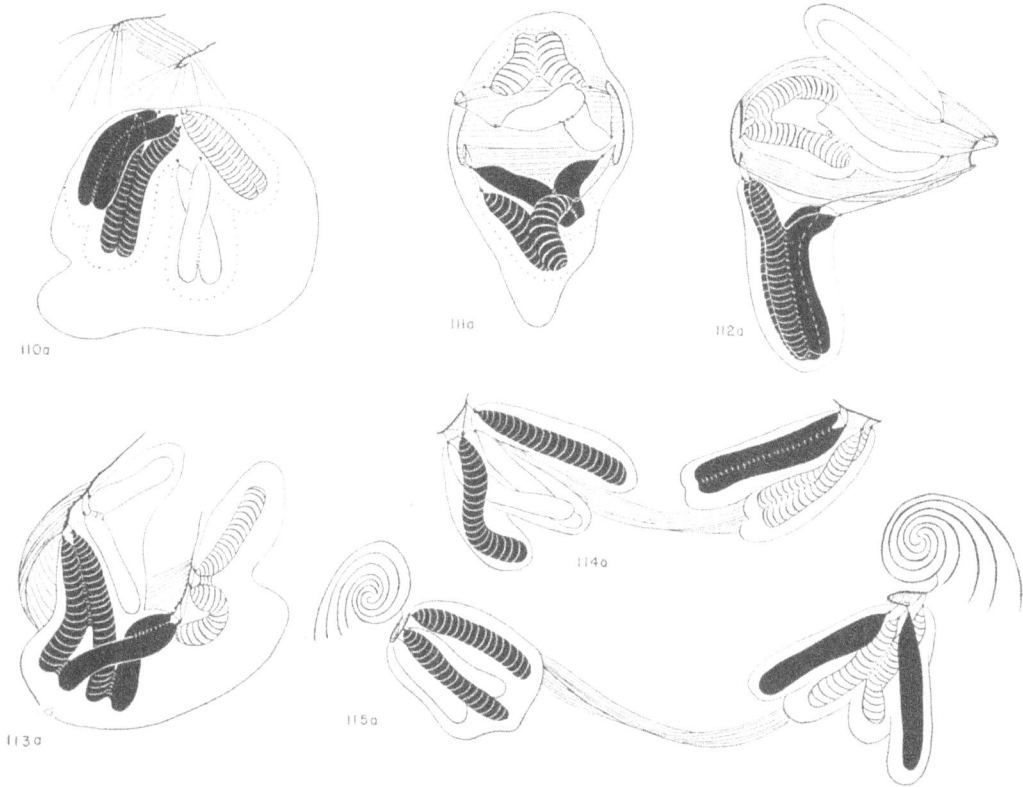

FIGS. 110a–115a. Same as figs. 110–115 except half chromatids are omitted in order to show more clearly the relation of chromatids to chromosomes in pairing, tetrad formation, chiasmata, and segregation. Even though half chromatids are not shown in these figures, the chromatids drawn must be referred to as half-sisters instead of sisters, if confusion is to be avoided. Proximal ends of centrioles omitted in most of the drawings. Free astral rays omitted in all figures except 110a. × 1,200.

110a. Pairing of half-sister chromatids is complete in some chromosomes and nearly so in others. No pairing of chromosomes.

111a. The paired half-sister chromatids of each chromosome are going to opposite poles. No chromosome pairing. Segregation of chromatids in each chromosome.

112a. The half-sister chromatids of each of the four chromosomes are going to the same pole. No segregation. Note how the tetrad (below, dark) is opening up; poleward movement is the same as it would have to be if the half-sister chromatids of each chromosome had a common centromere, as in meiosis I in most organisms where this process has been studied. Note also that half-sister chromatids of this tetrad are paired from end to end, another feature closely resembling the usual type of meiosis I. No interlocked chiasmata.

113a. A slightly later stage in the opening up of a tetrad, formed by the long (dark) chromosomes, than shown in fig. 112a. And this tetrad, unlike that of the previous figure, shows an interlocked chiasma; at centromere ends half-sister chromatids are paired, while at distal ends homologous, non-half-sister chromatids are paired. However, one can see clearly that the chiasma will be resolved without crossing-over (there will be no segregation of any portion of half-sister chromatids). Pairing at centromere ends in half-sister short (light) chromatids means there was no segregation in these chromatids unless crossing-over occurred.

114a. Pairing from one end to the other of half-sister chromatids in daughter nucleus on right means there was no segregation.

115a. Pairing of half-sister, short (light) chromatids in daughter nucleus on right suggests there was no segregation.

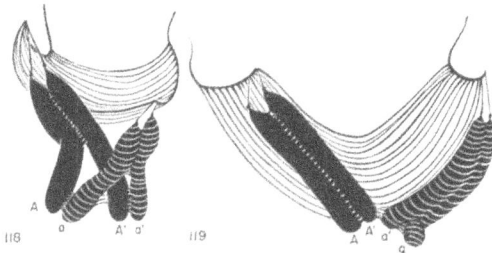

FIGS. 118 and 119. *Holomastigotoides diversa.* Diploid race with single chromatids. × 1,150.

118. Early anaphase. Even though the pairing of non-sister chromatids is restricted to a short region at the distal ends (and it was probably no more extensive earlier), it was sufficient to prevent segregation of sister chromatids.

119. Late anaphase. This figure shows that a very small amount of non-sister chromatid pairing (there may have been a little more earlier) was sufficient to prevent segregation of sister chromatids.

situation, as one does in mitosis, by regarding these meiotic chromatids as half chromatids, for, unlike the mitotic half chromatids, these chromatids are the product of chromosomal duplication. However, irrespective of the terminological difficulties—which I do not propose to settle at this time—haploid organisms with double chromatids present the same situation from the standpoint of segregation that one sees in meiosis I, namely, the sister half chromatids at mitosis cannot go to separate poles because they have a common centromere, nor can the sister chromatids at meiosis for the same reason. In both cases, crossing-over serves as a partial means of overcoming the embargo or ban which a common centromere places on segregation.

In diploid organisms at meiosis I there is no segregation of sister chromatids unless crossing-over occurs; there is segregation of homologous chromosomes. The sister chromatids (now chromosomes) segregate at II, some time after centromere duplication. But in diploid organisms at mitosis the possibilities of segregation are greater; because, unlike meiosis I, each chromatid has its own centromere during mitotic nuclear division, thus making it possible, under certain conditions, for segregation to occur in sister chromatids.

Mitosis is considered in two types of diploids in this paper. One type has single chromatids, the other double. Both have the same number of terminal centromeres in a given stage or 8 after duplication, which occurs in early prophase. In order to shorten and to simplify the description somewhat segregation (of chromatids and chromosomes) is considered in both types at the same time. No confusion will arise in doing this if the reader will remember that the paired chromatids in the race with single chromatids are sisters and those in the race with double chromatids are half-sisters. Also, in each instance when figures are cited, the num-

bers from 99 through 117 always refer to the double chromatid race.

In both single and double chromatid types, as we have seen, there is pairing of sister chromatids in the first type and of half-sisters in the second, as well as pairing of homologous non-sister chromatids, which is the same as chromosome pairing (figs. 89a–94a, 99a–110a). If we leave crossing-over out of consideration for the present, we note segregation of paired sister (and half-sister) chromatids in some instances and not in others (figs. 94a–98a, 111a–113a). In other words, two paired homologous chromosomes, each with two centromeres and two chromatids, may go, intact, to opposite poles (figs. 95a, 98a, 112a, 118–119), as is always the case in meiosis I, or the two homologous chromosomes may not pair at all, in which case there is segregation of their sister chromatids (long, dark ones only in figs. 94a, 97a, and both long and short ones in fig. 111a).

Such observations bring up the question, does pairing of chromosomes, however slight it may be, always cause each chromosome to act, in poleward movement, as a unit, and thus prevent the sister chromatids from going to opposite poles? Does the mere joining, for example, of homologous chromosomes, as in the short (light) ones of figure 94a or the long (dark) ones of figure 98a, prevent segregation in the chromatids of these chromosomes? Let's have a look at several specific examples and determine under what conditions the pairing of non-sister homologous chromatids affects the segregation of sister chromatids. In the short (light) chromosomes of figure 94a, the sister chromatids of each chromosome are paired from one end to the other, while there is only a slight amount of pairing of non-sister chromatids (and this is restricted to their distal end), yet there is plainly no segregation of sister chromatids. In the long (dark) chromosomes of figure 113a, sister chromatids (in each chromosome) are paired near their centromere ends and non-sisters at their distal ends, with a non-paired region in the middle. Here, as in figure 94a, there is clearly no segregation of sister chromatids (of the short chromosomes). In the long (dark) chromatids of figure 91a there is about the same amount of pairing of non-sister chromatids as in the short chromatids of figure 94a, but this pairing, unlike that of figure 94a, is at the centromere ends, and we can see, from the manner in which the centromeres are connected to the poles, that there will be either complete segregation in the chromatids of both of these chromosomes, or there will be crossing-over in two of the non-sister chromatids. A similar situation, but in a later stage, is seen in figure 92a where crossing-over is in progress.

If we turn our attention to a more detailed consideration of segregation of chromatids in those instances where there is no pairing of chromosomes (the homologous chromosomes either lie a considerable distance apart and in separate portions of the developing new nuclear membrane or each homologue has its own portion of the new membrane and pairing in either case is impossible), we note that, in every instance, the two paired chromatids go to opposite poles. For example, see: long (dark) chromatids, 93a, 94a, 97a; short (light) chromatids, figures 91a, 92a, 98a; and both long and short chromatids, figure 111a. But why shouldn't these paired chromatids go to opposite poles? They always do in haploid organisms with single chromatids (figs. 67a, 69a, 70a, 72a) or with double chromatids (fig. 82a). And, of course, in haploids and diploids they do likewise when they are not paired (fig. 68a). Pairing, then, clearly plays no role in poleward movement or segregation of chromatids in haploid organisms. However, these paired chromatids are sisters, and in the diploid examples just cited (figs. 91a–94a, 97a, 98a, 111a) the chromatids are also either sisters or half-sisters. This, then, narrows the problem to the point where we may ask the specific question, does the pairing of homologous non-sister chromatids affect the segregation of paired sister chromatids? If we were to look only at the paired short chromatids of figures 94a and 95a, or the paired long ones of figures 95a and 112a, we would conclude that non-sister chromatid pairing always prevents segregation in sisters. But there is no pairing, in any of these examples, of non-sister chromatids at, or even near, the centromere ends. On the other hand, if we look at figures 91a, 92a, 101a, and 107a, where the non-sister pairing of chromatids is at the centromere ends, we see clearly that there will be segregation (= crossing-over) of a portion of the sister chromatids near the centromere ends unless the pairing of non-sister chromatids in this region is resolved. It probably will not be, certainly not in figure 92a, and probably not in the others since poleward movement places no strain or tension on the pairing of non-sister chromatids. But the non-sister chromatid pairing at the distal ends of the chromatids, such as shown in the short chromatids of figure 94a and in both short and long chromatids of figure 95a, will be resolved during poleward movement and without segregation of chromatids. Resolution of non-sister chromatid pairing at distal ends—and without segregation— is in progress in figure 113a. In these long, straight rod-shaped chromatids, pairing of non-sister chromatids which extends for more than half their length may be resolved provided only sister chromatids are paired at the centromeres (fig. 112a). Of course, the possibility of segregation (= crossing-over) in one or more regions is far greater in median or submedian chromatids at meiosis I where pairing of sister and non-sister chromatids begins much earlier, when the chromatids are longer and where, as pairing progresses, a good deal of relational twisting of chromosomes is produced.

If confusion is to be avoided, one must not fail, as Darlington (1935) does, to differentiate between relational coiling of chromatids and relational twisting of chromosomes. He says,

In such a rare configuration the first chromosome, which has evidently neither paired nor crossed over in its intercalary region, is seen to have its chromatids coiled round one another. This is relational coiling of chromatids (fig. 6; cf. Newton and Darlington, 1929, pl. III, fig. 8d).

In Crepis with genetical suppression of chiasma formation, the diplotene chromosomes show strong relational coiling. In chromosomes or parts of chromosomes with normal chiasma formation, on the other hand, coiling at diplotene is usually slight and always uniform in distribution (Richardson, 1935a).

These observations indicate that chiasmata replace coiling of chromosomes and of chromatids. Coiling is normally seen at pachytene, but it survives to a full extent at diplotene only when crossing-over has failed through genetical suppression or through lack of pairing in intercalary regions. In order to see how the coiling can be replaced by chiasmata, we must consider first how the coiling itself arises. In pairing, the chromosomes come together side by side without any tendencious abnormality, without any predetermined spiral relationship, since they may come together at different points at the same time, and according as their chance dispositions in the nucleus allow. Coiling develops after pairing (Darlington, 1935). In mitosis the chromosomes must divide so that their daughter chromatids lie parallel, since in ring chromosomes they are capable of separating freely.

Coiled relationships of threads, whether chromosomes or chromatids, are therefore peculiar to the paired condition. How do they come about? When the lateral attraction between the chromosomes lapses at diplotene, the separated chromosomes uncoil. They have therefore been lying in a state of stress maintained by their lateral attraction acting in opposition to their longitudinal tension. The lateral attraction, to do this, must have positional specificity; the genes must be incapable of relative rotation after pairing.

Relational coiling of chromatids, as clearly set forth in my 1949a paper, is produced by the duplication of a coiled chromosome, while relational twisting of chromosomes is produced by the twisting about each other of two chromosomes which earlier were entirely free— in fact, they may have been some distance apart. In zygotic meiosis, they are—prior to fertilization—not only in separate nuclei, but are often in separate gametes (Cleveland, 1949b, figs. 56–125).

Also the fact that the sister chromatids of most meiotic chromosomes have one or more gyres of relational coiling as pairing begins, as well as after it is completed (photos. 170–173), serves to produce partial segregation (= crossing-over) much more frequently in such meiotic chromatids than in the mitotic ones of either of the diploid races of H. diversa.

Now, if we look at several of these same figures again —but from a different point of view—we note in 92a, 101a, and 107a in the region near the centromeres where the non-sister chromatid pairing occurs that there is no pairing, in this region, of sister chromatids. Hence, it is obvious in these figures, particularly in 92a, that poleward movement places no strain at all on the non-sister pairing. The non-sisters of each chromosome in this region go to the poles as a unit—as a chromosome. In so doing, however, they do place a strain on the pairing of sister chromatids in the rest of each chromosome. And as a result of this strain, one of two things happens: (1) sister chromatid pairing is resolved and segregation of sister chromatids occurs; or (2) both sister and non-sister pairing remain intact (no pairing is resolved) during poleward movement. But this cannot happen unless the tension resulting from the pairing of sister chromatids in one region and non-sisters in another is relieved in some way during poleward movement This, as we shall see in the next section, is relieved by segregation (= crossing-over) of a portion of the two inner, interlocked, non-sister chromatids. On the other hand, if we look at figure 113a from the same point of view, it is obvious that the pairing of non-sister chromatids at the distal ends of the long (dark) chromosomes, may be resolved without either total or partial segregation. Also the situation in the long (dark) chromatids of figure 112a may be resolved in the same manner so far as the centromere ends are concerned, for at these ends only sister chromatids are paired, and in each case the sisters are behaving as a unit in poleward movement. Whether partial segregation occurs or not in some other region depends on which type of pairing is resolved (one must be), namely, that between the sister chromatids or that between the non-sisters, during poleward movement. However, if both sister and non-sister chromatids were to pair from one end to the other—something that never occurs in *H. diversa*—then sister and non-sister chromatid pairing could not be differentiated, and it would thus be impossible to determine the role of chromosomal pairing in segregation.

Now, we may conclude in the two diploid races of *H. diversa* (from the examples of nuclear division illustrated and from over 200 studied but not illustrated) that: (1) there is always segregation of entire sister chromatids when pairing occurs in sisters only; (2) when pairing occurs in non-sister chromatids anywhere except at the centromere ends, together with pairing of sisters from one end to the other, there is never segregation of entire sister chromatids, but there is partial segregation in those few instances where crossing-over occurs; (3) when non-sister chromatids only are paired at centromere ends and sisters only elsewhere, partial segregation (= crossing-over) occurs in all cases except those where terminalization resolves the pairing of sister chromatids and thus permits complete segregation; (4) when both sister and non-sister chromatids are paired at centromeres and sisters only elsewhere—a type of pairing that has been seen only once—segregation of sister chromatids probably occurs.

Thus, we see clearly the function of chromosomal

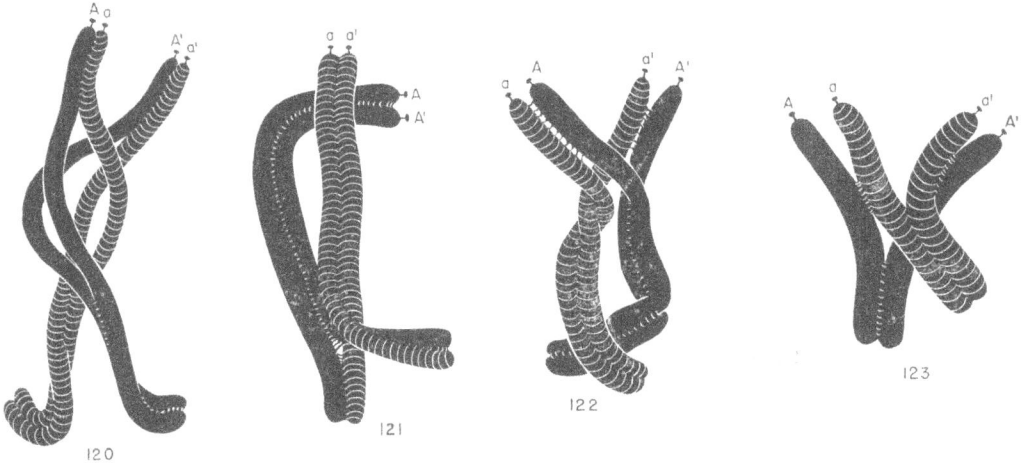

FIGS. 120-123. *Holomastigotoides diversa*. Prophase tetrads in which only sister chromatids are paired in one region and only non-sisters in another, thus showing how interlocked chiasmata are produced. (If sister chromatids pair from one end to the other before pairing of non-sisters begins, then the pairing of non-sisters in one or more regions produces chiasmata, but in such chiasmata the non-sister chromatids are not interlocked. See figs. 95a, 106a, 109a, 112a, 153, and photo. 189.) × 1,400.

120. Chromatids still fairly long: non-sisters are paired at centromere ends in each chromosome, sisters are paired at distal ends in each chromosome, and there is no pairing for a considerable distance between these regions.

121. In this tetrad, there are two regions where pairing of chromatids occurs: non-sisters are paired at distal ends and sisters along the same chromosomes for the rest of their length. No unpaired region.

122. Sister chromatids of each chromosome are paired for distal two-thirds of their length, and non-sisters are paired for a considerable distance near their centromere ends.

123. Sister chromatids are paired distally: those of the chromosome on the right are paired for approximately half their length, while those of the other chromosome are paired for a shorter distance. Pairing of non-sister chromatids is less extensive (no pairing near centromere ends).

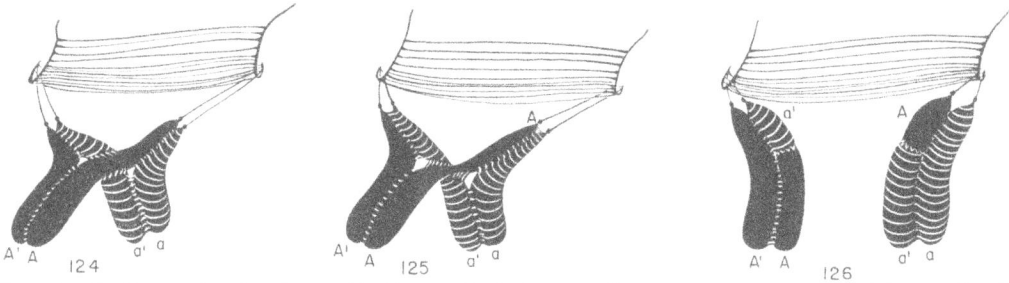

FIGS. 124–126. *Holomastigotoides diversa*. Diploid race with single chromatids. Three stages or steps in partial segregation of sister chromatids or crossing–over, resulting from the pairing in the prophase before poleward movement began of sister chromatids in one region and non-sisters in another. The last stage is diagrammatic. × 1,500.

124. For one-third of their length, beginning at the centromere end, non-sister chromatids are intimately paired. For the rest of the length of the chromatids, sisters are paired (there is no region where both sister and non-sister chromatids are paired). Pairing of this type produces an interlocking chiasma at the point where non-sister chromatids A and a' cross each other. From centromere ends to a point of interlocking, there is no tension on the pairing between non-sisters. Hence, this pairing remains intact during poleward movement. The tension is on the pairing between the sisters A'A and a'a and on the non-sisters only at the point where they are interlocked. As a result of this tension, which is produced by poleward movement, one of two things happens. The first is the pairing of the sister chromatids of each chromosome is pulled apart (resolved), thus bringing about complete segregation of sister chromatids

and centromeres, something that could not happen in meiosis I, owing to the fact that sister chromatids at this division have a common centromere, and therefore could not pair at their centromere ends as these mitotic chromatids are doing. The second thing that may happen is for the pairing of the sister chromatids of each chromosome to remain intact, as well as that between the non-sisters which is not under tension, and for the interlocked non-sisters to pull almost in two (fig. 125) at their points of interlocking, but shortly before they are pulled in two union of the nearly free ends begins, so that no ends ever become entirely free. Then, while union is in progress, the remaining strands connecting centromere and distal ends of chromatids A and a' pull apart, to be followed shortly by union, thus resulting in a partial segregation or crossing-over of sister chromatids (fig. 126).

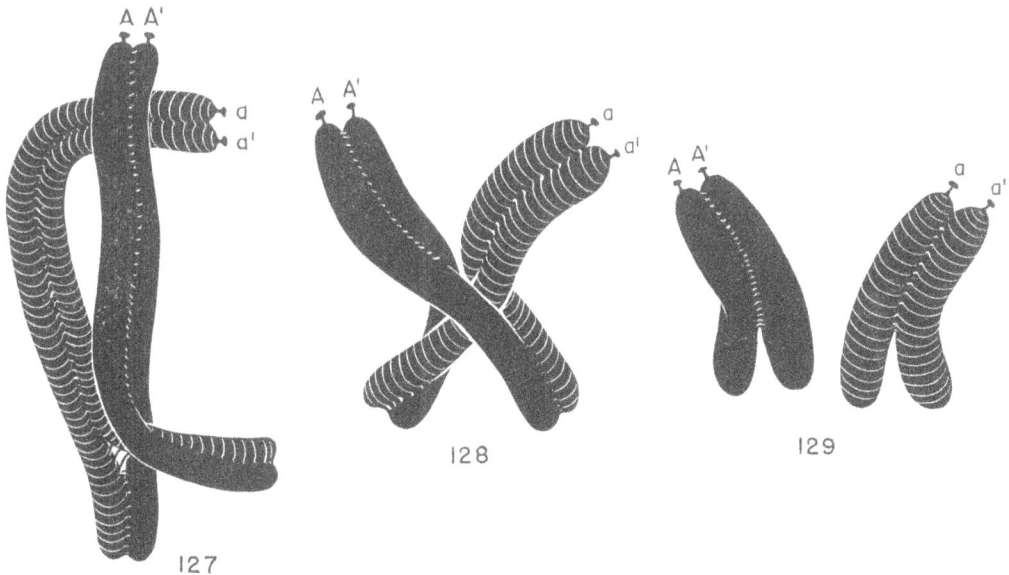

FIGS. 127–129. *Holomastigotoides diversa*. Non-segregation of chromatids and the resolution of an interlocked chiasma by terminalization. This is the usual occurrence when such chiasmata occur near the distal ends of chromatids in mitosis (if near the centromere ends as in fig. 92a, it is not). There is no strain during poleward movement on the pairing of sister chromatids of either chromosome; hence this pairing remains, while that of non-sister chromatids, which is under strain, is removed (figs. 128, 129). × 1,850.

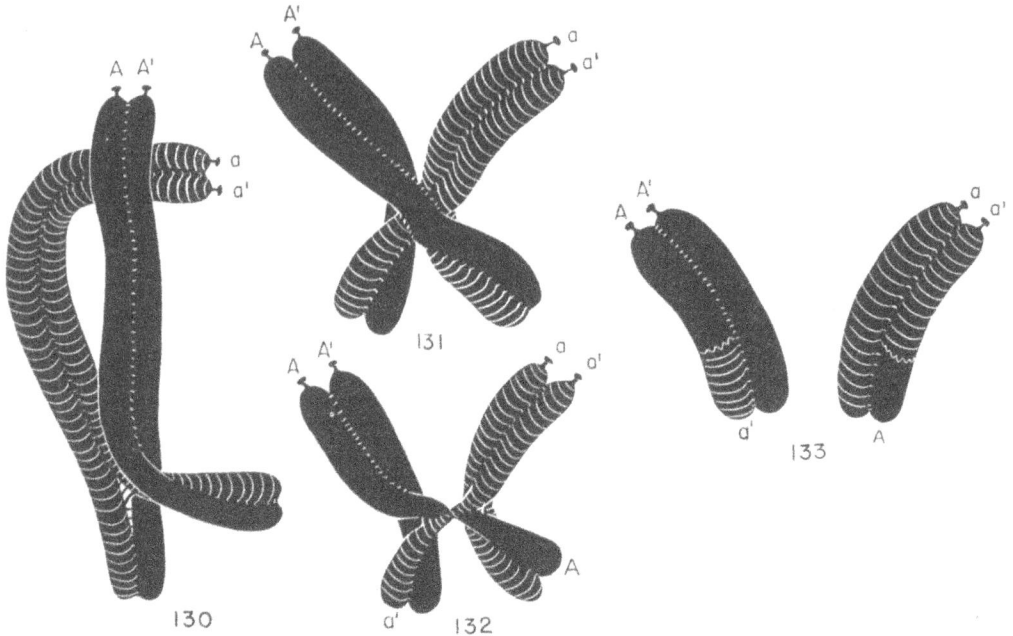

FIGS. 130–133. *Holomastigotoides diversa.* Chiasma not resolved. Since pairing of neither sister nor non-sister chromatids is resolved, interlocking of non-sister chromatids remains as homologous chromosomes move apart at their centromere ends (fig. 131). As poleward movement continues, tension narrows two of the chromatids—one from one chromosome and one from the other—at their common point of interlocking, the point where pairing changes from that of sisters to non-sisters (fig. 132). However, before these interlocked chromatids are pulled in two, union between the ends of non-sisters begins, so there are never any free ends of chromatids (fig. 132). As the centromere ends of the homologous chromosomes move farther apart, pulling in two and union are completed, and thus two precisely comparable segments of non-sister chromatids cross over (fig. 133). × 1,850.

pairing in mitosis is to prevent segregation. However, owing to variation in the manner in which the pairing occurs, it sometimes fails partly, and at other times altogether, to accomplish its mission.

Next, what is accomplished by chromosomal pairing during meiosis? Since sister chromatids, of meiosis I (owing to failure of centromeres to duplicate themselves), have a common centromere, segregation is already inhibited by this change in behavior from that seen in mitosis, so that pairing can only serve as an additional check on the segregation of entire chromatids. We thus see a clear-cut difference in the segregation of chromatids in mitosis and meiosis: In mitosis, there is always complete or entire segregation of sisters in the absence of pairing, but in meiosis I this is made impossible by the fact that sisters have a common centromere. Then, pairing of homologous chromosomes at meiosis I has only one function: it makes partial segregation or crossing-over possible.

CROSSING–OVER

If my conception of the manner in which crossing-over and pairing are interrelated is correct, the underlying mechanism of crossing-over has already been explained fairly well in the sections on pairing and segregation. On the other hand, if crossing-over is not related to pairing, then my conception of the process breaks down completely, and another must be sought. But the fact that crossing-over is always associated with pairing makes me feel that my observations should be considered, along with the observations and concepts of other investigators, for whatever they may be worth.

There are some workers who feel certain there is a direct relationship between chiasmata and crossing-over, and of course chiasmata—regardless of how they may be formed—can only be seen when pairing occurs. However, these workers feel that chiasmata are produced at the time of pairing only because crossing-over has already occurred, i.e., before pairing began (figs. 142, 144). In other words, a chiasma occurs wherever there has been a cross-over between non-sister chromatids.

Let's let Darlington (1935) state this point of view.

First let us consider the exact time of crossing-over. It must occur before the chiasmata appear, since it determines them. It cannot occur before the chromosomes divide into chromatids, since the independent change of the chromatids shows that this division is complete at the time of crossing-over. But this division is observed to occur, in conformity with the movements of the chromosomes observed at all other times, at the moment they begin to repel one another. Crossing-over must therefore be simultaneous with division, repulsion and chiasma formation.

Note a few lines above the contradictory statement "It [i.e., crossing-over] must occur before the chiasmata appear, since it determines them." But Kaufmann (1934), Cooper (1948, 1949) and others have clearly shown that chiasmata occur in the absence of crossing-over, just as Sax (1930, 1932) earlier anticipated.

Now, if crossing-over results from the manner in which the four homologous chromatids pair—as I think it does—then a chiasma may or may not produce a cross-over, depending on whether there is terminalization or not. By this I mean the pairing between non-sister chromatids may be resolved (figs. 127–129), in which case crossing-over does not occur, or it may not be resolved, in which case crossing-over does occur (figs. 130–133).

In most material where cytological observations have been made on chiasmata, the chromatids cannot be seen at all until the paired chromosomes have begun to "open up." Indeed, many workers, since chromatids cannot be seen earlier, think only of chromosomal pairing. In fact, they think chromatids are not present when chromosomes pair, but are produced, by chromosomal duplication, after pairing has been established. Yet, they think crossing-over of chromatids occurs before the chromosomes pair, at which time, according to their own state-

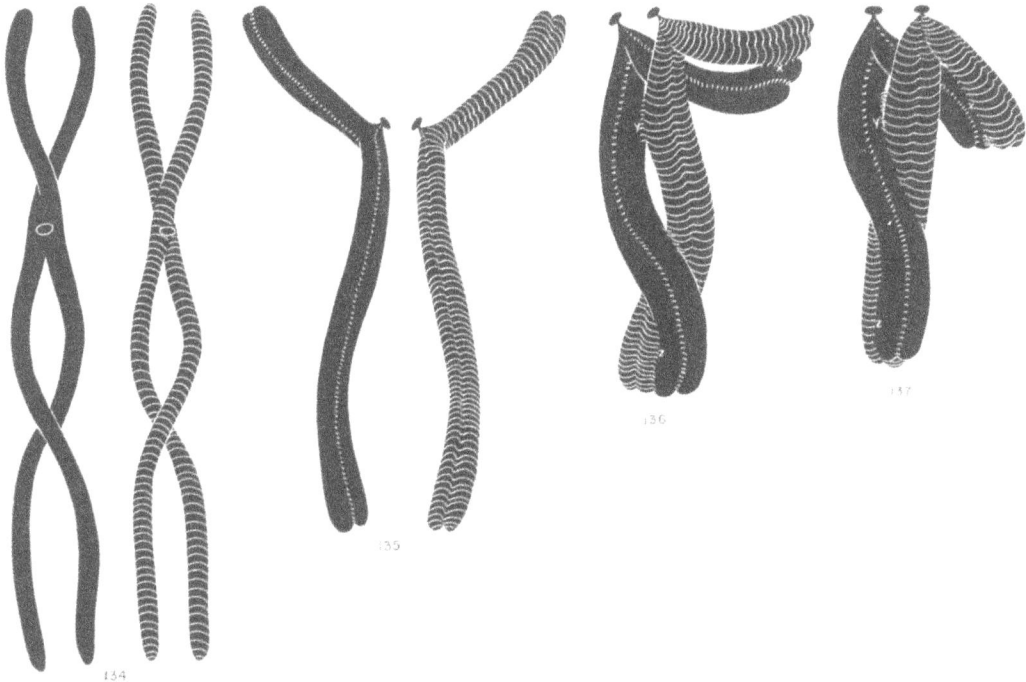

FIGS. 134–137. *Barbulanympha ufalula.* Pairing of prophase chromatids at meiosis I in such a manner that no interlocked chiasmata are produced.

134. Two homologous chromosomes. There is a slight amount of relational coiling in the sister chromatids of each chromosome.

135. The relational coiling has nearly all come out and the sister chromatids of each chromosome have paired from one end to the other. Sister chromatids have a common subterminal centromere.

136. The sister chromatids of each chromosome are still paired from one end to the other and, in addition to this, pairing has occurred between non-sister chromatids in three regions, *x, y,* and *z* (this has brought the homologous chromosomes together and, therefore, may be regarded as chromosome pairing).

137. Similar to previous figure except later stage. The tetrad appears shorter and the pairing of non-sister chromatids is more extensive. Three non-interlocked chiasmata have been produced by the pairing of non-sister chromatids in regions *x, y,* and *z.* As the chromosomes move to opposite poles these regions may be resolved (pull apart) without crossing-over or they may not be resolved and crossing-over may occur, depending on whether the pairing of sister or non-sister chromatids is stronger in a given region. The sister chromatids, except in those regions where cross-overs occur, must go to the same pole because they have a common centromere.

ments, no chromatids exist. How, then, can there be crossing-over in something which does not exist?

Not all arguments for crossing-over prior to pairing are so easily refuted as this one. If we admit four chromatids are present—even though in most organisms we cannot see them—before pairing begins—and we must do so because this is a *sine qua non* of crossing-over—there is the possibility that non-sisters pair in certain regions and sisters in other regions, and thus produce chiasmata, because crossing-over occurred earlier. What is the evidence against this? One thing is the fact that the homologous chromosomes, in many forms, lie a considerable distance away from each other prior to the time of their pairing. Other chromosomes

often lie between the homologues. Facts such as these offer almost insurmountable obstacles to the idea that pieces of chromatids—sometimes very short ones indeed—travel by themselves over considerable distances, through a highly granular non-chromatin nuclear environment (as revealed by phase contrast observations on living cells during division), and yet manage to reach the proper place on the proper chromatid before the homologous chromosomes come together and begin to pair. If these difficulties are not sufficient to make such a mechanism of crossing-over appear almost impossible, there is the additional fact that the crossing-over piece of a chromatid from one homologous chromosome is always precisely identical in length and composition with

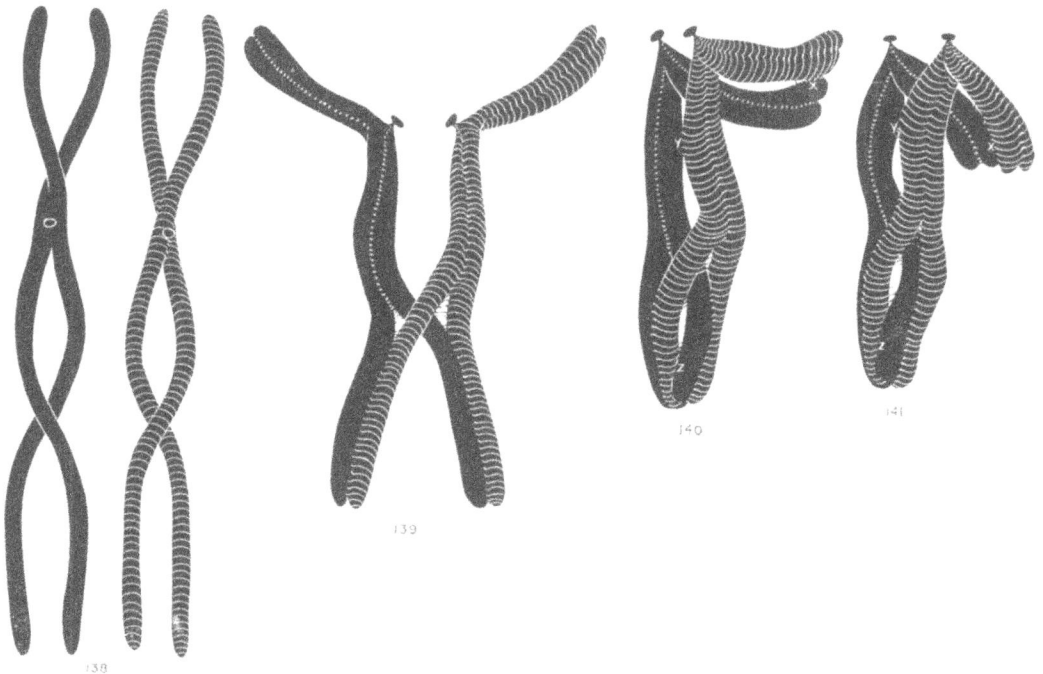

Figs. 138–141. *Barbulanympha ufalula.* Pairing of prophase chromatids at meiosis I in such a manner that an interlocked chiasma is produced.

138. Two homologous chromosomes with a small amount of relational coiling of sister chromatids.

139. The relational coiling of sister chromatids has nearly all come out (sometimes more remains at this stage) and there is pairing of chromatids. But this pairing here, unlike that of fig. 135, is of two kinds. Sister chromatids are paired for a considerable distance on both sides (up and down) of their common centromere, while in the distal third of each chromosome non-sisters are paired. This type of pairing produces a chiasma in which the two inner, non-sister chromatids are interlocked at exactly the same point on each chromatid. (Since the pairing of non-sister chromatids has brought the two homologous chromosomes together, it is the same thing as chromosomal pairing.)

140. Later prophase in which the same pairing of the previous figure remains and in addition to this pairing of non-sister chro-

matids is beginning in three other regions, one on either side of the centromeres and one near the distal ends (*x*, *y*, and *z*).

141. Late prophase, almost metaphase. The pairing of non-sister chromatids in the regions *x* and *z* is more extensive, while that of region *y* remains as well as the pairing of sister and non-sister chromatids shown in fig. 139. In this tetrad, then, there are two kinds or types of chiasmata. One is produced earlier than the other (fig. 139) by non-sister chromatid pairing in a region before sisters (sisters only in one or more regions and non-sisters only in one or more regions). In the other or second type of chiasma, sister chromatids pair in the region of the chiasma before non-sisters. This second type, as a rule, is more easily resolved as chromosomes move in opposite directions beginning with their centromeres.

that from the other chromosome. And there is also the further fact that sometimes several cross-overs occur simultaneously in the same two chromatids. How could these chromatids maintain a semblance of their continuity while each is being broken into many short pieces which float or move freely through space before joining up again? Certainly a simpler and less haphazardous method must be responsible for crossing-over. This one, it seems to me, would produce mostly deletions—if, indeed, a cell could survive it. Sometimes there is no homology expressed in the union of fragments produced by spontaneous aberrations (figs. 41, 45, 48–51). Why then should there always be homology in crossing-over if this process results from the transfer of fragments prior to pairing?

There are other arguments against crossing-over occurring prior to pairing, but I shall not discuss them because, in my opinion, less arguments and more facts are needed if we are ever to understand the mechanism of crossing-over. Nor shall I give further consideration, at this time, to a general or historical account of crossing-over. A brief consideration of such material may be found in most text-books dealing with cytology, together with references to more detailed articles.

Let's consider briefly what we can see clearly during mitosis of the diploid races of *Holomastigotoides diversa* and during meiosis of *Barbulanympha*. In the first

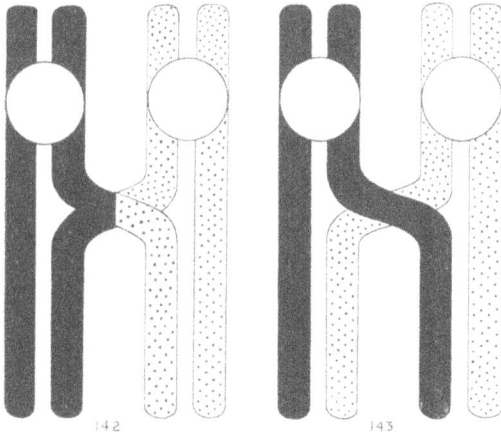

Fig. 144. A conventional diagram (see De Robertis *et al.*, 225, White, 83) of two chiasmata resulting from two cross-overs between two non-sister chromatids, the cross-overs having occurred before the chromatids began to pair.

Fig. 145. Same as fig. 144 except crossing-over has not occurred. In this diagram two interlocked chiasmata have been produced by the pairing of non-sister chromatids only in the mid-region of each chromosome and of sisters only at centromere and distal ends of each chromosome. Both chiasmata may be resolved without crossing-over, one may be resolved and the other not, or neither may be resolved.

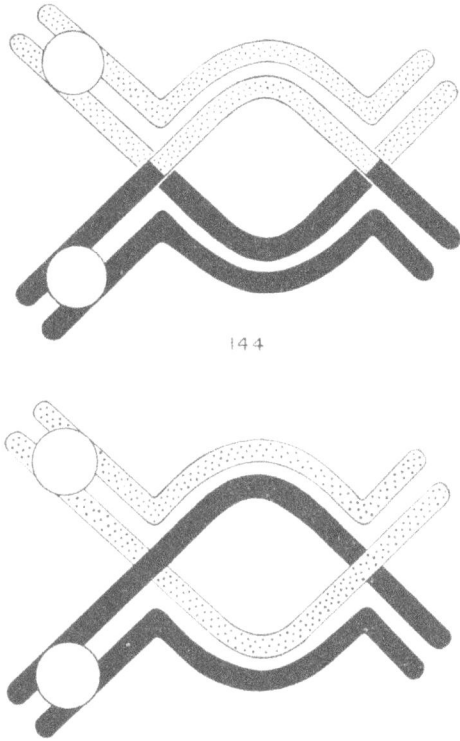

Fig. 142. A conventional diagram (see De Robertis *et al.*, 223) of a single chiasma resulting from one cross-over between two non-sister chromatids. In other words, the crossing-over occurred before the chromatids began to pair.

Fig. 143. Same as fig. 142 except crossing-over has not occurred. In this diagram an interlocked chiasma has been produced by the pairing of sister chromatids only in the upper half and of non-sisters only in the lower half of each chromosome. If this chiasma is resolved later, as the chromosomes separate, there will be no crossing-over. If it is not resolved, there will be crossing-over of the two inner, non-sister chromatids at the point where they are interlocked.

place, there is no stage in the entire life cycle of the chromosomes, beginning with chromosomal duplication in the very early prophase, when chromatids cannot be seen plainly; indeed, they may be photographed in the living cell (photos. 169–173, 190–192). Further, the nature of the pairing in *H. diversa* is always such (as explained earlier in this paper) that the pairing of sister chromatids can be differentiated from that of non-sisters. Now, since the behavior of chromatids can be followed with unmistakable clarity from the time of chromosomal duplication through all the events of pairing and poleward movement, one can see the initial stages in the pairing of sister and non-sister chromatids, and in those organisms with half chromatids one can also see very plainly the initial stages in the pairing of sister and non-sister half chromatids (figs. 101, 107). In diploid mitosis, sometimes only sister chromatids pair, and of course no chiasmata are formed, but in meiosis of *Barbulanympha* there is always pairing of sister and non-sister chromatids, leading to the formation of chiasmata (figs. 136, 137, 139–141; photos. 173, 190–192).

In both mitotic and meiotic pairing, two kinds of chiasmata are produced. If non-sister chromatids pair first in one or more regions and sisters elsewhere, the non-sisters become interlocked at the point or points where the chromatid pairing changes from that of sisters to non-sisters (figs. 92a, 107a, 113a, 120–123, 124–126, 130–133, 139–141). On the other hand, if sister chromatids pair first from one end to the other before pairing of non-sisters begins, chiasmata are produced, but the non-sister chromatids are not interlocked (figs. 95a, 106a, 109a, 134–137).

Perhaps those workers who see chiasmata without crossing-over are dealing mostly with this type of chiasma, while those who see a close correlation between chiasmata and crossing-over are dealing mainly with the other type where the non-sister chromatids are interlocked. If so, the contention of each group is correct.

In order to avoid duplication, the actual manner in which crossing-over occurs—as I see it—between interlocked non-sister chromatids is considered in the description of the figures (124–126, 130–133), and therefore need not be restated in the text.

DISCUSSION

Since no need has been found for such cumbrous terms as pachytene, leptotene, diplotene, and zygotene, they have not been used. Clearer and simpler explanations may be made without them.

If the terms reductional division and equational division are dropped, we will experience less difficulty in explaining meiosis to beginning students. However, I shall not be disappointed if such a reform in terminology is a long time coming, because cytologists, as well as other biologists, enjoy using these terms too much to give them up without a struggle, however desirable it may be. Of course, since meiosis means reducing, the terms meiosis I and meiosis II, if applied only to chromosomes, are equally awkward and confusing. However, if we state that centromeres are reduced in I and chromosomes in II, then these terms become less confusing. But what shall we use in place of equational division? We can simply say there is no segregation of entire sister chromatids in meiosis I, although there may be partial segregation (= crossing-over). In meiosis II, unless there was crossing-over in I, the sister chromatids, which should now be regarded as chromosomes, undergo complete segregation.

The present tendency to replace the older term tetrad by bivalent is confusing and in my opinion should be discontinued. I also fail to see a need for the term univalent. An unpaired chromosome is more accurate and less confusing, because often a so-called univalent is composed of two chromatids.

If crossing-over occurs before chromatids pair, why doesn't it occur between sisters as well as non-sisters? Cross-overs do not occur in sister chromatids because these go to the same pole in meiosis (and in mitosis, too, when there is pairing of chromosomes) and hence are under no strain or tension, but the pairing of non-sister chromatids is placed under tension during poleward movement because they are anchored to opposite poles, and, as a consequence, are pulled almost in two at the point (or points) where the pairing changes from that of sisters to non-sisters. However, shortly before they are pulled in two, union between non-sisters begins and thus prevents a complete break from occurring in either chromatid and, at the same time, relieves the tension (figs. 131–133).

Certain workers have felt for some time that in the pairing of chromosomes the homologues (= non-sister chromatids) are held together only at certain points and that each point represents a chiasma or region where crossing-over occurred earlier. However, direct observations on the pairing of living chromosomes refute this notion so completely that there is no point in discussing it. It is perfectly clear that the non-sister chromatids come together and pair for the very same reason that sisters do, namely, because they attract each other. It is just as ridiculous, then, to argue that the non-sister chromatids in a diploid come together and pair because of previous cross-overs as it would be to argue that sister chromatids in a haploid come together and pair for the same reason.

Although Sax (1930, 1932), in my opinion, came nearer than any one else grasping the true mechanism of crossing-over, he did not realize there is pairing of sister chromatids in one region and non-sisters in another. He thought, as did other workers later, that each point of contact between homologues represented a permanent exchange of partners rather than the pairing of non-sister chromatids. In several of his figures there are paired and unpaired regions of chromatids, although he did not interpret them in this manner. His so-called "split" regions, in all probability, represent unpaired regions, and his "undivided" regions represent paired ones.

Pairing of chromatids can change from that of sisters to non-sisters only so often along a tetrad. This is the factor which determines the distance between cross-overs, the so-called phenomenon of interference.

SUMMARY

Asexual reproduction has been studied in two haploid ($n = 2$) and two diploid races of the hypermastigote flagellate *Holomastigotoides diversa*. The haploid races differ from each other in that one has single and the other double chromatids, although there are two centromeres in each race. The diploid races differ in the same way. This closely related series of independent races makes an ideal group for studying pairing, segregation, and crossing-over. Pairing, for example, may be studied in sister half chromatids, in half-sister chromatids, and in homologous chromosomes, all of which may be seen with diagrammatic clarity throughout their entire life cycles.

Pairing of sister chromatids in single-chromatid haploids appears to accomplish nothing. It certainly has no effect on the segregation of chromatids. Likewise, in diploids with single chromatids the pairing of sister chromatids *only* during mitosis has no effect on segregation. Each sister invariably goes to a different pole. In other words, there is segregation (or crossing-over) of entire chromatids. But when there is pairing of homologous non-sister chromatids as well as sisters, segregation of entire sister chromatids is rare, although partial segregation, which is the same as crossing-over, occurs fairly often. Thus the usual accomplishment of chromosomal pairing (= pairing of non-sister chromatids) in mitosis is to prevent complete segregation of sister chromatids and to make possible, at certain times (depending on the manner in which chromatids pair), crossing-over.

Zygotic meiosis has been studied in the hypermastigote flagellate *Barbulanympha* in which both sister and non-sister chromatids are diagrammatically plain. Since sister chromatids at meiosis I have a common centromere, complete segregation in these chromatids is thus made impossible, irrespective of whether the homologues pair or not. At most, pairing of homologues can only serve as an additional check on segregation of chromatids. But no such check is needed. Then, the pairing of homologous chromosomes at meiosis I served only to make partial segregation or crossing-over possible. And we may conclude that mitotic chromosomal pairing, except in rare instances, suppresses complete segregation of chromatids and makes possible partial segregation, while meiotic pairing merely makes partial segregation or crossing-over of chromatids possible. The inhibition of centromere duplication in meiosis I and not in mitosis is responsible for the difference in pairing function in the two processes.

The exact manner in which the pairing of sister and non-sister chromatids makes crossing-over possible cannot be summarized briefly.

References will be found at the end of the fourth paper of this series. I am greatly indebted to Mrs. Laura O. Samuelson for the interest and care she has taken in making the drawings.

IV. PHOTOMICROGRAPHS OF LIVING CELLS DURING MEIOTIC DIVISIONS

Everything considered, there is no better organism than *Barbulanympha* for studying division in the living cell. Some hypermastigotes with only two large chromosomes are better for structural detail of chromosomes, but no organism is as good for centrioles, centrosomes, and achromatic figure as *Barbulanympha*, and its centromeres are nearly as plain as those of any cell. A few of those hypermastigotes in which only mitosis occurs show individual chromatids slightly better, but no organism known shows chromatids and chromosomes so clearly during meiosis as *Barbulanympha*. Nucleoli, of which there are many, are also plain. These statements result from many years of observation of living cells, both before and after phase contrast equipment became available.

The photomicrographs were made with special light and dark phase contrast apparatus prepared for me by Mr. L. V. Foster of the Scientific Department of the Bausch and Lomb Optical Company. The light source in most instances was the Zirconarc, the film was Kodak M plates, and the camera was Bausch and Lomb L Model.

Barbulanympha, along with many other genera of hypermastigote flagellates, lives in the hind-gut of the wood-feeding roach *Cryptocercus punctulatus.* It, like all of its relations, develops a sexual cycle only when its host molts. The meiotic phase of the sexual cycle of *Barbulanympha* occurs 18–24 hours after molting when the temperature is approximately 70° F. The insects were observed many times daily and when one was seen in the process of molting or still white after molting, it was isolated in a metal or glass container and examined 18–24 hours later. The best preparations are obtained by taking a small, undiluted drop of fluid from that portion of the hind-gut adjacent to the iliac valve. A cover glass is placed on the drop immediately. The drop should be just large enough to fill evenly all the space between slide and cover glass, with no fluid left over. Since the insect does not feed during molting there will seldom be any wood particles in the preparation, but if there are they should be removed or the preparation discarded. At first, such preparations were compressed in a compression chamber, but this produced so much distortion in these large cells and their organelles that the process was soon discontinued.

Three species of *Barbulanympha* are included in the photomicrographs, but more are of the two larger ones, *B. ufalula* and *B. laurabuda.* However, in the smaller species, *B. estaboga,* which is shown in figures 154, 155, 176, one can usually get more organelles—such as flagellated areas, centrioles, achromatic figure, nuclear membrane, and chromosomes—in focus at the same time. The magnification is stated approximately in connection with the description of each figure.

Only a few remarks regarding the morphology of *Barbulanympha* are needed in order to understand the photomicrographs (for a detailed description see Cleveland *et al.,* 1934). A comprehensive account of meiosis in this genus is given in a separate paper in preparation.

The two unusually large resting stage centrioles, with centrosomes 5 to 6 microns in diameter surrounding their distal ends, are the most striking feature of this organism (figs. 146, 147). They reproduce themselves and, in addition, produce the achromatic figure, centrosomes, flagella, parabasals, axostyles, nuclear sleeve, and cap. The flagellated areas, of which there are two, are restricted to a small region at the anterior end of the cell. They may be seen in several of the photomicrographs. Since no attempt was made to include parabasals and axostyles in the pictures, further mention of them at this time is unnecessary.

In most cases it was impossible to focus sharply on both chromosomes and the achromatic figure at the same time because so often they lie at different focal levels (fig. 186 is best for both). Consequently, some of the pictures—such as figures 150–168—show the achromatic figure and centrioles better than they do the chromosomes because the camera was focussed sharply on these organelles instead of the chromosomes. Of course, the chromosomes are fairly plain in a few of these figures, because they happened to be in nearly the same focal plane as the achromatic figure and the centrioles. On the other hand, in figures 169–173 and 178, 179, the camera was focussed on chromosomes, and, whenever portions of the centrioles and the achromatic figure show, it is only because they lie in nearly the same focal level as the chromosomes.

Since the chromosomes themselves lie at so many focal levels, it is impossible to show very many of those in a nucleus in a single photograph. And of course the same applies to the centromeres. The entire number may be seen very plainly—plainer than after any method of fixation and staining—but, if centromeres are to show plainly, one should focus on only one or two of them at a time (compare figs. 163, 181, 182, 186 with 164; in 163, 181, 182, 186 several may be seen, but not plainly; in 164 and 189 two may be seen plainly. Figures 163 and 164 should be viewed after 90° clockwise rotation).

In the resting cell, the distal ends of the centrioles produce no astral rays (fig. 148); and in the prophase, the duplication of chromosomes occurs some time (fig. 149) before the production of astral rays begins. However, it is not long after the centrioles begin to produce astral rays before those from one centriole meet, join, and grow along those from the other, and thus form the very early central spindle portion of the achromatic figure (figs. 150, 151). Presently, as the joined and overlapping astral rays increase in length, the central spindle becomes longer (fig. 152) and approaches the nucleus (fig. 153). Soon, the astral rays begin to extend over the anterior border of the nucleus (figs. 154, 155). As the central spindle becomes longer, it begins to depress the nuclear membrane (figs. 156, 157), finally depressing it so much that, when viewed from a high focal level, it appears to have cut the membrane into two portions (fig. 159); but, when viewed from a mid-focal level, it is seen merely to have depressed the membrane to the centre (figs. 160, 161, 165, 167, 182).

It is not long after the central spindle begins to depress the nuclear membrane before some of the free astral rays that extend over the nucleus join the centromeres which lie in the nuclear membrane and thus become chromosomal fibres (figs. 163–165). Then, as the central spindle elongates, owing to growth and increase in length of the astral rays of which it is composed, the chromosomes of meiosis I and II are slowly moved apart (figs. 160, 162, 165, 177, 182, 185). As the growth process continues, the central spindle becomes longer, and thus finally separates the chromosomes into two distinct groups or daughter nuclei (figs. 174, 185), each group being surrounded by a portion of the parent nuclear membrane which is reorganized during late telophase without ever being broken down completely (central portion of parent membrane in fig. 185 is discarded).

The picture one obtains of the distal end of a centriole depends on where the camera is focussed. If the focus is the upper level, one sees the centriole extending into a round centrosome from which the astral rays appear to radiate (figs. 153, 160). But if one focusses on the centre of the centrosome instead of its upper surface, one can see the central spindle fibres leading directly into the centriole, which narrows somewhat as it terminates in the centre of the centrosome (figs. 151, 152, 154, 155, 159, 161). One can also see in several of the photomicrographs the free astral rays arising from the distal end of the centriole and passing through the centrosome (figs. 154, 155 are best).

Several photomicrographs have been made of the achromatic figures when more than two centrioles are present, but since these figures and the centrioles which produce them occupy so much space, particularly up and down, one either has to use low magnification, which shows very little detail, or else include only a portion of them (figs. 166–168).

The relationally coiled sister chromatids in meiosis I may be seen soon after duplication occurs and some time before the centrioles begin to produce astral rays (fig. 149). As the relationally coiled sister chromatids unwind, they become plainer (fig. 169). As the homologous prophase chromosomes come together preparatory to pairing, one to three gyres of relational coiling of their chromatids still remain intact (fig. 170). This relational coiling of sister chromatids remains throughout the pairing process. As the homologues pair, they become slightly twisted about each other (fig. 171), the amount of twisting (not coiling) varying considerably in the

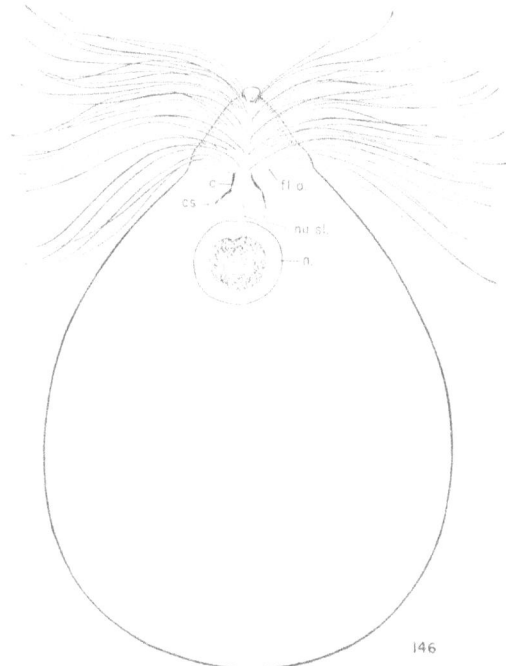

FIG. 146. *Barbulanympha.* Entire cell in resting stage, showing a surface view of the two small, flagellated areas (*fl. a.*) at the anterior end, the nuclear sleeve (*nu. sl.*), nucleus (*n*), centrioles (*c*) and centrosomes (*cs*) surrounding the distal ends of the centrioles. × 260.

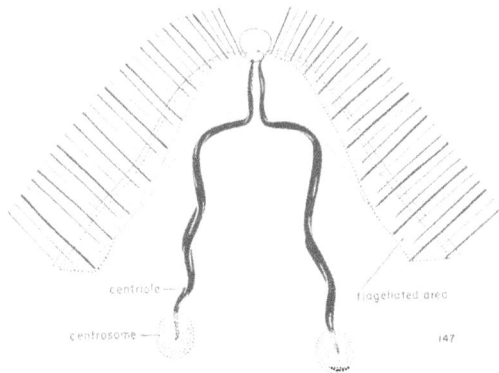

FIG. 147. *Barbulanympha.* Detail of the anterior end of a resting cell. × 2,000.

same homologues at different times. The chromatids at this time have many gyres of tight major coils, which may be seen when the exposure time is exactly right (fig. 172). One can also see pairing of sister and nonsister chromatids (fig. 172). In fact, the chromosomal pairing is initiated by the pairing of non-sister chromatids (fig. 171). Finally as sister and non-sister chromatids pair in more regions, tetrads are produced (figs. 173, 190–192) and, in many cases, even in the photomicrographs, each of the four chromatids can be seen clearly. Chromosomes in the last stages of tetrad formation bend on themselves at their centromeres. In a tetrad then we have: (1) bending of chromosomes in such a manner that the two arms of the same chromosome often lie near each other; (2) relational twisting of homologous chromosomes resulting from the difference in length of homologues as they pair; and (3) a small amount of relational coiling of sister chromatids that was produced at the time of chromosomal duplication and that did not come out during the unwinding period following duplication. It is highly important, if we wish to obtain a clear and complete picture of chromosomal behavior, especially during pairing and poleward movement, to differentiate between relational coiling of sister chromatids, relational twisting of homologous chromosomes, and the bending of individual chromosomes on themselves.

In the resting stage between meiosis I and II, minor coils may be seen fairly well (fig. 178). In the late prophase of meiosis II the paired and pairing chromosomes (they were chromatids in I) may be seen plainly (fig. 179). In anaphase II chromosomes appear longer than in I (compare figs. 181, 182). They appear longer in anaphase II because their major coils are looser (fig. 183). In late telophase II major coils are still farther apart and many chromosomes, as well as centromeres, lie in the same focal plane (figs. 187, 188).

DISCUSSION AND CONCLUSIONS

The following rather general statements are based only in part on the photomicrographs of *Barbulanympha* presented here. Nuclear division has been studied, both in living and fixed cells, in twelve other genera of hypermastigotes with the same care as in *Barbulanympha*, and a considerable amount of time has also been devoted to a study of the living cells of higher organisms during division.

Most biologists—cytologists included—use the word "spindle" in much the same way that the layman uses the word "stomach"; sometimes it refers to a portion of the alimentary canal, while at other times it refers to the entire alimentary canal, or even to the abdomen. Likewise, "spindle" as used by most biologists sometimes refers to a portion of the achromatic figure; at other times it refers to all of the achromatic figure. A dual usage of this sort sometimes promotes misunderstanding. However, it is probably just as difficult—at least in this instance—to get the biologist to use precise terms as the layman, and I shall not attempt it.

The term "blepharoplast," for which there was never any need, is also used in two ways. Some biologists,

particularly protozoologists, write "blepharoplast" when in reality they mean centriole; at other times, they write "blepharoplast" when the organelle they have in mind is a basal granule.

Wilson's (1925) attempt to evade the necessity of unravelling the long-standing centrosome-centriole controversy by coining the term "central apparatus" accomplished nothing except to delay for several years the broad, intense study of these organelles needed for an understanding of them. Many centrioles have no centrosomes surrounding them, while others are either partly or entirely surrounded by centrosomes. But if one wishes to write precisely—in fact, correctly—the two terms cannot be used as synonyms. An achromatic figure is always produced by centrioles—never by centrosomes.

The inappropriate and altogether unnecessary terms "paradesmose" and "intradesmose" are sometimes used —mostly by a few protozoologists—for the central spindle portion of the achromatic figure in certain protozoa where the nuclear membrane remains intact during most of nuclear division. A so-called paradesmose, in reality, is an extranuclear central spindle and an intradesmose (sometimes written centrodesmose) is an intranuclear one. However, it must be perfectly obvious, if one takes the time to consider the situation, that a central spindle is a central spindle, irrespective of whether the nuclear membrane remains intact or disappears. The difference in behavior is in the nuclear membrane—not in that of the central spindle or any other portion of the achromatic figure.

Since the central spindle portion of the achromatic figure in all mitoses and meioses is formed by astral rays meeting, joining, and growing along one another, the term "anastral mitosis" therefore becomes meaningless and should be dropped. Another and even more valid reason for retirement of the term "anastral mitosis" is the fact that free astral rays become chromosomal fibres, and thus are responsible for the movement of the chromosomes to the poles. However, unless the future text-book writers are more wide awake than the present-day ones, it will be many years before students are not required to learn this empty term.

The orderly distribution of chromosomes at the time of nuclear division is so important for the continuous function and existence of a cell that many attempts have been made to understand and to analyze the responsible mechanism. However, these attempts, for the most part, have ended in the formulation of theories which, in many ways, contradict one another.

On a percentage basis, more time perhaps has been wasted arguing over the origin and nature of the achromatic figure than any other cellular entity in animals or plants, and nearly as high a percentage of time has been wasted on speculation regarding its function. But I do not propose to repeat—or even to consider—any of these arguments, because the *Barbulanympha* and other material with which I have worked is so perfectly clear that to do so would be as ridiculous as to argue whether oak trees bear leaves or not.

Without centrioles, no achromatic figure is produced, and without an achromatic figure, there is no nuclear division or poleward movement of chromosomes. When a sister chromatid in mitosis is attached to a centriole, via chromosomal fibre and centromere, there is no attraction whatever between this centriole or pole and the other sister chromatid. The same is true of homologues at meiosis I. But the chromatid, or the homologue, which loses its attraction for a pole when its mate becomes attached to this pole, does not lose its attraction for any other of several poles which are sometimes present. It can go to any of them, provided they do not lie too far away for an astral ray to make contact with its centromere.

REFERENCES

CLEVELAND, L. R. 1949a. The whole life cycle of chromosomes and their coiling systems. *Trans. Amer. Philos. Soc.* **39**: 1–100.

——. 1949b. Hormone-induced sexual cycles of flagellates. I. Gametogenesis, fertilization, and meiosis in *Trichonympha*. *Jour. Morph.* **85**: 197–296.

CLEVELAND, L. R., S. R. HALL, E. P. SANDERS, and J. COLLIER. 1934. The wood-feeding roach Cryptocercus, its protozoa and the symbiosis between protozoa and roach. *Mem. Amer. Acad. Arts and Sci.* **17**: 185–342.

COOPER, K. W. 1948. The evidence for long specific attractive forces during the somatic pairing of dipteran chromosomes. *Jour. Exp. Zool.* **108**: 327–336.

——. 1949. The cytogenetics of meiosis in Drosophila. Mitotic and meiotic autosomal chiasmata without crossing over in the male. *Jour. Morph.* **84**: 81–121.

DARLINGTON, C. D. 1935. The time, place and action of crossing-over. *Jour. Genet.* **31**: 185–212.

——. 1937. Recent advances in cytology. 2nd ed. London, Churchill.

DEROBERTIS, E. D. P., W. W. NOWINSKI, and F. A. SAEZ. 1948. General cytology. Phila., Saunders.

KAUFMANN, B. P. 1934. Somatic mitoses of *Drosophila melanogaster*. *Jour. Morph.* **56**: 125–155.

SAX, K. 1930. Chromosome structure and the mechanism of crossing over. *Jour. Arnold Arb.* **11**: 193–219.

——. 1932. The cytological mechanism of crossing over. *Jour. Arnold Arb.* **13**: 180–212.

SHARP, L. W. 1934. Introduction to cytology. N. Y., McGraw-Hill.

WHITE, M. J. D. 1945. Animal cytology and evolution. Cambridge Univ. Press.

WILSON, E. B. 1925. The cell in development and heredity. 3rd ed. N. Y., Macmillan.

852

EXPLANATION OF PLATES

All photomicrographs are of *Barbulanympha*. The approximate magnification is stated in connection with the explanation of each figure. Except for figure 148 which is the resting stage before meiosis I, figure 178 which is resting stage between meiosis I and II, and figures 179, 182, 183, 186–188 which are of meiosis II, all figures are of meiosis I. Except for figure 190, where two scratches were removed, neither negatives nor prints were retouched in any way whatever.

PLATE 1

FIG. 148. Resting nucleus and the posterior portions of the two elongate centrioles. Note large spherical centrosome surrounding distal end of each centriole. A nuclear sleeve lies between the centrioles. Note large number of nucleoli with clear areas surrounding them. Dark contrast. × 2,000.

FIG. 149. Early prophase. Anterior end of cell and more of the centrioles than in previous figure. One can see how the fibres composing a centriole are twisted about one another. The two flagellated areas show faintly. Chromosomes have been duplicated. Note relational coiling of their sister chromatids. No astral rays have been produced and nucleoli are still present. Dark contrast. × 1,500.

FIG. 150. Slightly later prophase. Nucleus is mostly out of focus. Note early central spindle extending between distal ends of centrioles. One can see how each centriole curves inward at almost a right angle as it approaches a flagellated area. Thus the two come close together anteriorly and extend side by side almost to the rounded cap at the upper border of the photograph. The flagellated areas, which move apart later, still lie beside each other as in the resting stage. Dark contrast. × 1,000.

FIG. 151. Same as previous figure except light contrast. × 1,000.

FIG. 152. Slightly later stage. Central spindle is longer and broader. Pressure on cell exerted by weight of cover glass has broken nuclear sleeve and has thus permitted the nucleus to move slightly posterior to its usual position when the sleeve remains intact. Nucleus is almost out of focus. However, note large nucleolus with definite, dark bodies within it. Portion of one flagellated area shows at upper right. Light contrast. × 1,000.

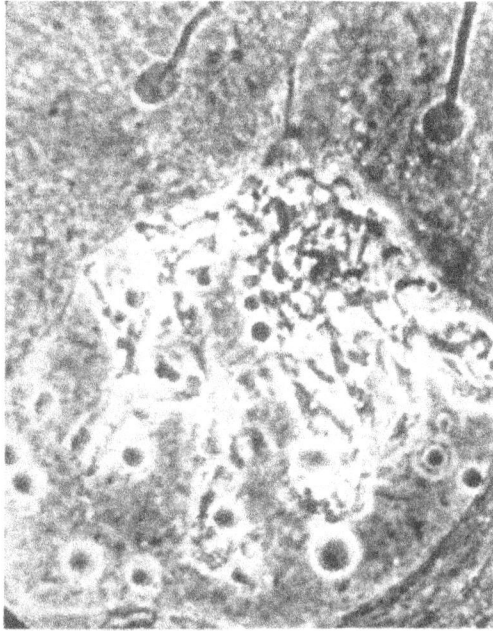

148

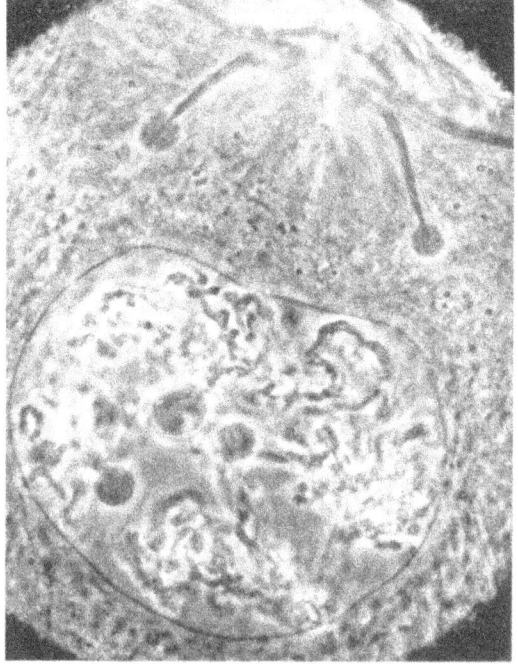

149

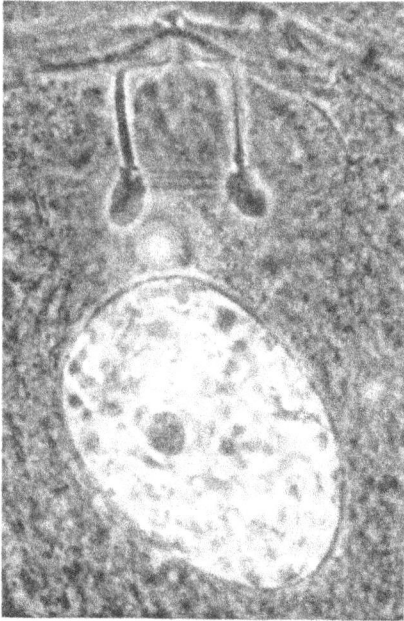

150

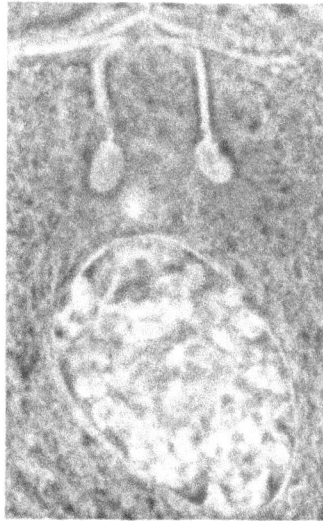

151

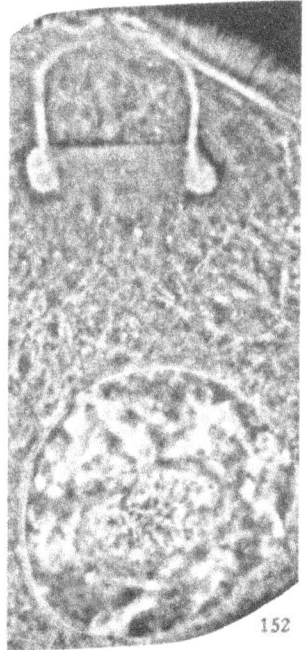

152

PLATE 2

Fig. 153. Slightly later prophase. Only the outline of the nuclear membrane shows faintly because nucleus is out of focus. The posterior half of each centriole shows plainly, as well as the centrosomes, and the early achromatic figure in which both free astral rays and those that have joined to form the central spindle show plainly. The anterior portions of the centrioles are lost underneath the flagellated areas. However, since these areas have not moved apart in the centre (no space between them as in figs. 154, 155), this achromatic figure is slightly earlier than that of fig. 152. Light contrast. × 1,400.

Fig. 154. Turn this figure, as well as the next one, clockwise 90° before viewing it. Later prophase. Lateral view of nucleus, achromatic figure, centrioles, and flagellated areas. Central spindle has just begun to depress the nuclear membrane and free astral rays are beginning to extend over its anterior surface. On the right, nearly all of the centriole and its flagellated area are plain. Note space between flagellated areas (they have separated). Chromosomes not in sharp focus. Light contrast. × 900.

Fig. 155. Only slightly later than previous figure and much like it in many ways. Central spindle has depressed nuclear membrane a little more and somewhat more of centrioles and their flagellated areas are in focus. Nucleus is not in sharp focus. However, one can see a tetrad slightly above centre on right and another near the lower left border. Light contrast. × 900.

Fig. 156. Slightly later stage and view is different. The flagellated areas and centrioles are viewed almost laterally while the view of the nucleus is nearly vertical. The central spindle is bending somewhat as it depresses the nuclear membrane. Chromosomes mostly out of focus. Light contrast. × 1,200.

Fig. 157. This is a considerably later stage viewed vertically to show how the larger and longer central spindle has depressed the nuclear membrane to the centre where it lies surrounded by faintly discernible chromosomes (note one centromere near lower margin of central spindle, slightly nearer its right hand end). Nuclear membrane is lobulated. All of centriole at right hand end of central spindle can be seen and most of the other one is visible. Flagellated areas, which now lie far apart, may be seen clearly. Note at right hand end of central spindle how the distal end of the centriole extends to the centre of the centrosome. Also note in this and previous figure many prominent cytoplasmic granules near the nucleus. These are always present in dividing cells from this stage onward. Light contrast. × 1,400.

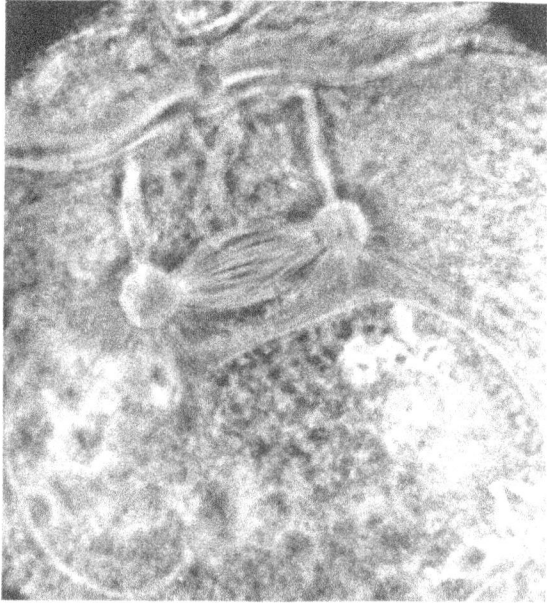

153

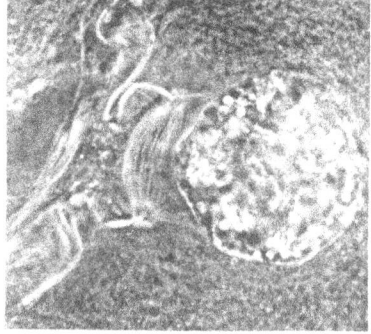

154

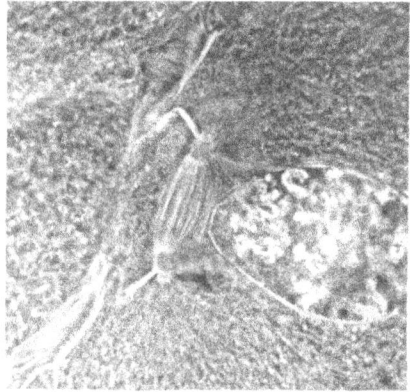

155

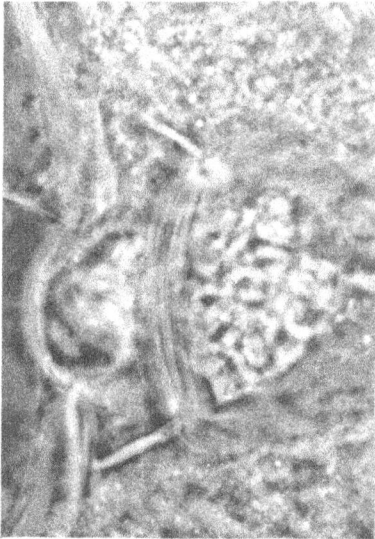

156

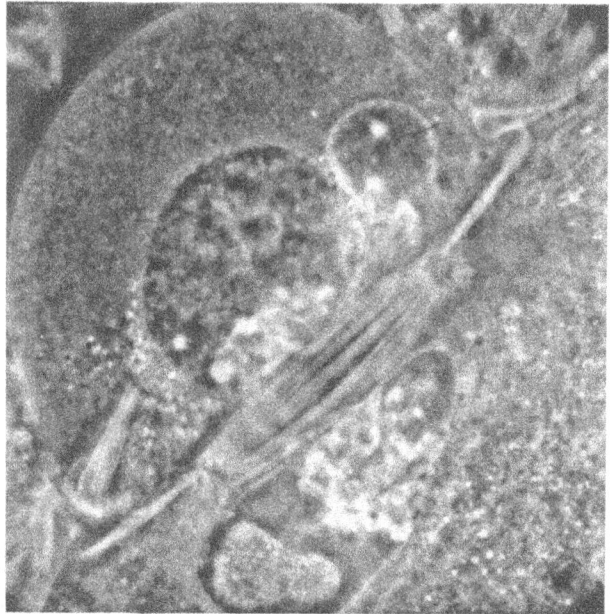

157

857

PLATE 3

FIG. 158. This is an earlier view than the previous figure of the central spindle depressing the nuclear membrane, and this view is lateral. The chromosomes are mostly out of focus and so is a good deal of the flagellated area on the left. However, most of each centriole and the compact central spindle are in focus and show plainly. Best view of this figure is after 90° counterclockwise rotation. Dark contrast. × 900.

FIG. 159. This figure is like 156 except slightly later. Flagellated areas, centrioles, and achromatic figure are viewed laterally while that of nucleus is almost vertical. Such a view helps one to understand how the flagellated areas move apart as the central spindle becomes longer and depresses the nuclear membrane. Here it appears to have cut the membrane in two portions, but it has not. A deeper focus would show it near centre of an intact nuclear membrane. Some free astral rays may be seen. Chromosomes mostly out of focus. Note two, small, new flagellated areas. One lies near anterior end of each centriole, between nuclear membrane and centriole, slightly nearer centriole. Light contrast. × 1,200.

FIG. 160. Vertical view of early anaphase. Central spindle has depressed nuclear membrane to centre and lies surrounded by chromosomes. Most of centriole and flagellated area on left in focus; only a portion of centriole on right in focus. Four centromeres may be seen, but they are not in sharp focus; they all lie below the central spindle—three on left and one on right. Light contrast. × 1,400.

FIG. 161. Centrioles lie 180° apart with the central spindle between them. Free astral rays show faintly. Chromosomes mostly out of focus. Dark lines near lower border and to right of centriole are axostyles. Dark contrast. × 1,400.

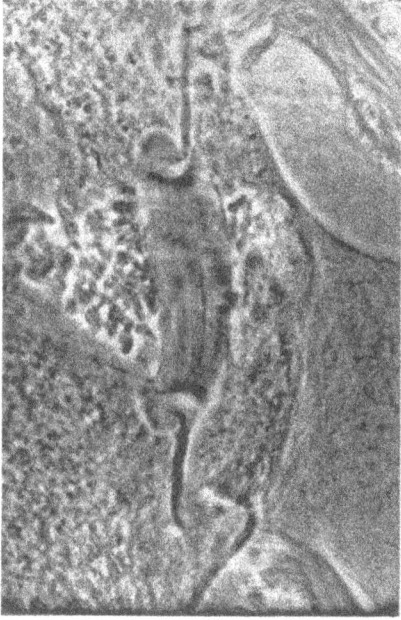

158

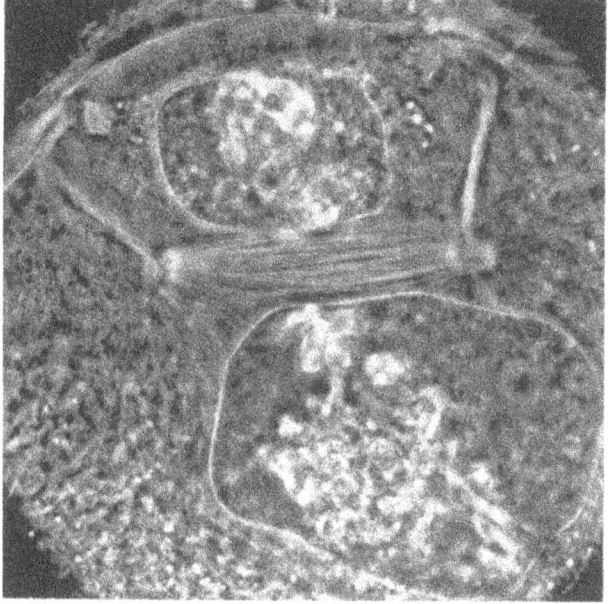

159

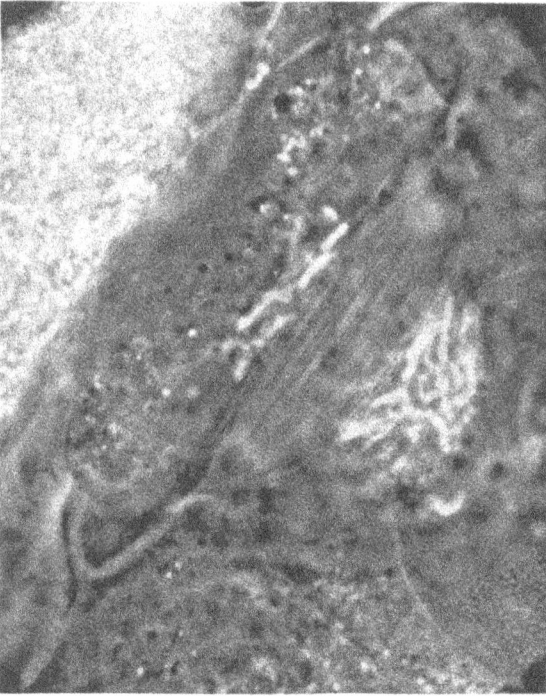

160

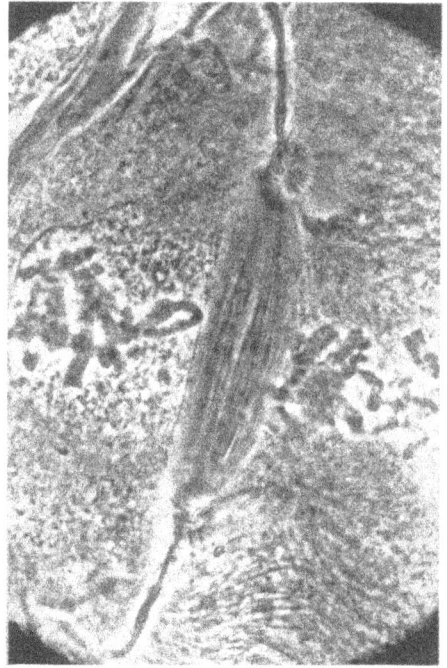

161

PLATE 4

FIG. 162. Same as figure 160 except dark contrast. Chromosomes are mostly out of focus. Centriole and flagellated area on left are plainer than in 160. × 1,400.

FIG. 163. This was an attempt to show several centromeres, but only two are at all plain (upper right, some distance above central spindle, lower left, fairly near central spindle, four others show faintly). Dark contrast. × 1,200.

FIG. 164. Rotate clockwise 90° before viewing this figure. The sole purpose of this photograph is to show two centromeres of one chromosome and their chromosomal fibers plainly, shortly after the time of centromere duplication (it occurs earlier in meiosis of *Barbulanympha* than in most organisms). They, of course, lie above central spindle to the left. Dark contrast. × 1,200.

FIG. 165. Early anaphase. Vertical view. Central spindle has depressed nuclear membrane to centre and centrioles lie 180° apart. This photograph is inserted primarily to show the nuclear membrane at this stage and how the centromeres are anchored to it (lie within the membrane). However, only two centromeres are in focus, and only one sharply so (upper right, some distance above central spindle). Light contrast. × 1,400.

FIG. 166. This low power photograph shows three central spindles fairly well, but the two centrioles on the right, which, like the two on the left, lie parallel to each other, can scarcely be seen at all (out of focus). The upper right centriole and the lower left one are each functioning in the production of a portion of two central spindles, while each of the other two centrioles is functioning in the production of a portion of a single central spindle (a single centriole when only two are present never produces more than one-half of a complete central spindle). Dark contrast. × 400.

FIG. 167. Here this low power photograph shows faintly how three centrioles, provided they lie at the corners of an imaginary equilateral triangle, can produce three central spindles, by each centriole doing double duty. This, and figs. 166, 168, produce final photographic proof—if any more proof is desired—that astral rays produce the central spindle. Light contrast. × 400.

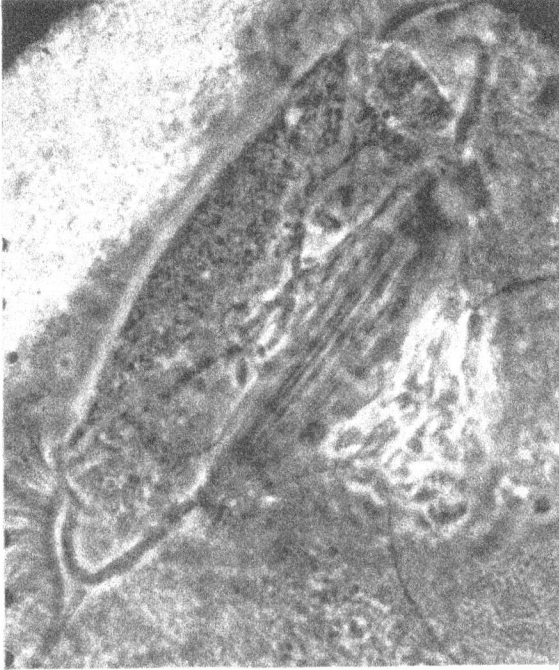

162

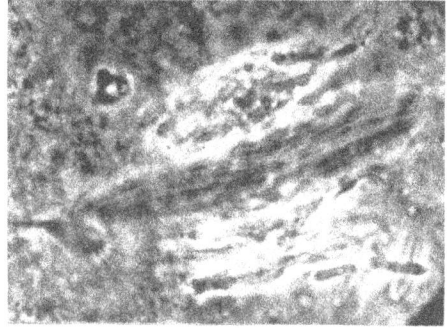

163

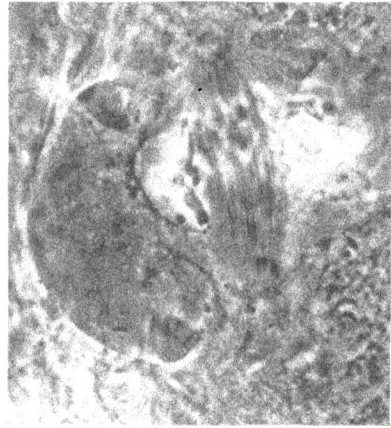

164

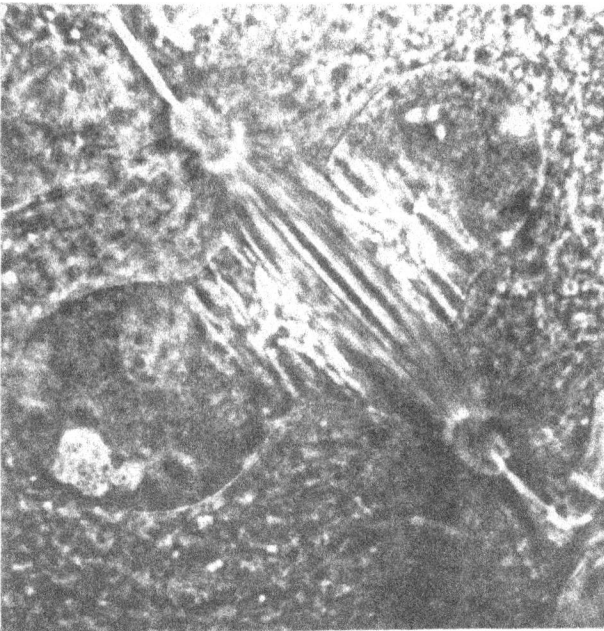

165

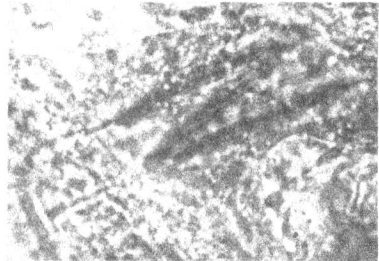

166

167

PLATE 5

FIG. 168. Multipolar achromatic figure produced by three centrioles, two above, which lie fairly close together, and one below. Chromosomes mostly out of focus. Dark contrast. × 1,400.

FIG. 169. Prophase I chromosomes shortly after duplication. Note relationally coiled sister chromatids. Centrioles and achromatic figure entirely out of focus. Dark contrast. × 1,400.

FIG. 170. Considerably later prophase I chromosomes. In upper centre two homologues have come together preparatory to pairing. Each homologue is composed of two sister chromatids. In the one on the right, about two gyres of sister chromatid relational coiling remain; the one on the left has slightly less relational coiling. The sister chromatids are pairing in some regions (note interconnections). One can see faintly individual major coils in each chromatid (a shorter exposure, as in fig. 172) would have shown them better). Centrioles and achromatic figure not in focus. Other homologues below are beginning to pair, but in no instance are entire homologues in focus. Dark contrast. × 1,400.

FIG. 171. Two prophase I homologous chromosomes slightly to right of centre are beginning to pair in two regions. Each homologue has two relationally coiled sister chromatids. There are about two gyres of relational coiling in the chromatids of each chromosome. Other homologues are beginning to pair, but in no instance are entire homologues in

focus. Centrioles and achromatic figure not in focus. Dark contrast. × 1,400.

FIG. 172. Note the four prophase I chromosomes which lie between the centrioles and which extend both above and below the central spindle. The two homologues which lie on the left are beginning to pair at their ends. Note particularly the strands connecting them at their upper ends. This is the beginning of non-sister chromatid pairing (= chromosomal pairing). In this region the sister chromatids are already paired for a considerable distance. At the opposite (lower) ends of these same homologues, one can see pairing of both sister and non-sister chromatids. Perhaps the two homologues on the right, in the region near a centriole, show tight, individual major coils of chromatids as plainly as anywhere in this photomicrograph. They also show pairing in sister chromatids plainly. The distal half of each centriole and part of the achromatic figure may be seen. Dark contrast. × 1,400.

FIG. 173. Late prophase I tetrads. In several instances, the sister chromatids of each homologue may be seen, as well as sister and non-sister chromatid pairing. Centromeres, which are median and submedian, lie at the points where the chromosomes bend on themselves, but not a single one is sharply in focus. A portion of one centriole and the central spindle may be seen to the right, slightly above centre. Dark contrast. × 1,400.

862

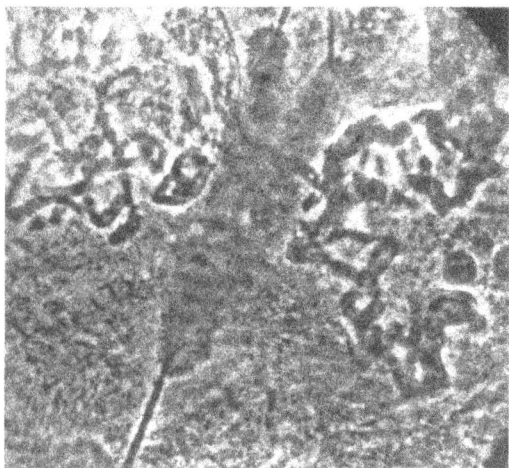
168

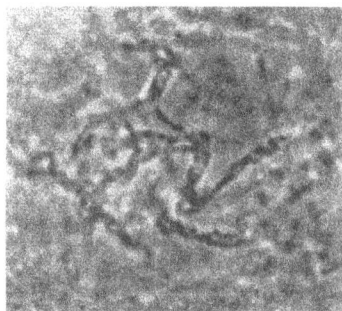
169

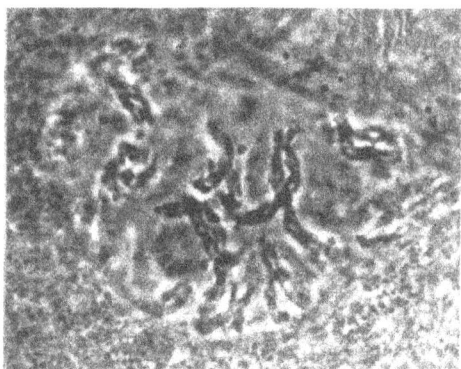
171

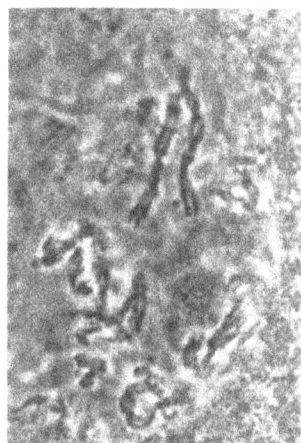
170

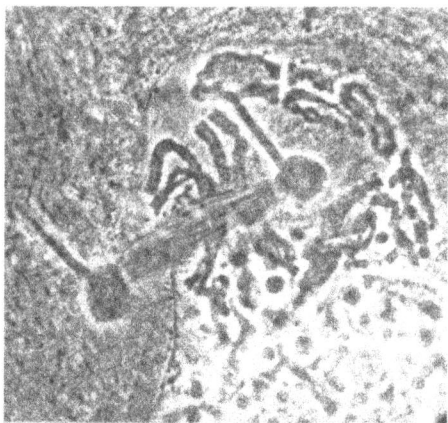
172

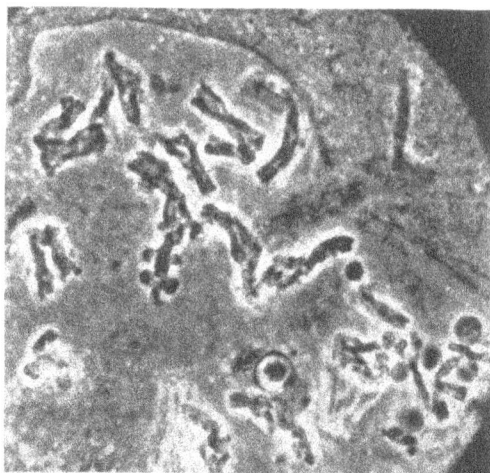
173

PLATE 6

FIG. 174. Telophase I (after nuclear division) with a long, straight central spindle extending between the daughter nuclei. Centrioles and centrosomes may be seen at each end of the central spindle. Nearly all of the lower centriole is in focus. The line beside it, and which bends with it, is the border of a flagellated area. Dark contrast. × 1,200.

FIG. 175. A rather poor attempt in one of the two large species of *Barbulanympha* to show prophase tetrads, centrioles, centrosomes, achromatic figure, and flagellated areas all at the same time. Light contrast. × 800.

FIG. 176. The same in the small species. The bare space between the flagellated areas is plain, and the centrioles are fairly good. Light contrast. × 800.

FIG. 177. A similar attempt at early anaphase to show all of these organelles at the same time. Flagellated area on left and the chromosomes perhaps show best. Remains of nucleolar material lie in the upper, left region of nuclear membrane. Light contrast. × 600.

FIG. 178. Resting stage between meiosis I and II. Centrioles and their centromeres not in focus. Borders of flagellated areas lie above nucleus. Dark contrast. × 1,500.

FIG. 179. Late prophase II, showing pairing of the chromosomes. Pair slightly below and to left of large, dark nucleolus perhaps shows best. Each chromosome is bent on itself at its centromere. There is no pairing in centromere regions. Note some conspicuous cytoplasmic granules near nucleus as in most figures of meiosis I and II but not in the resting stage between I and II (fig. 178). Dark contrast. × 1,400.

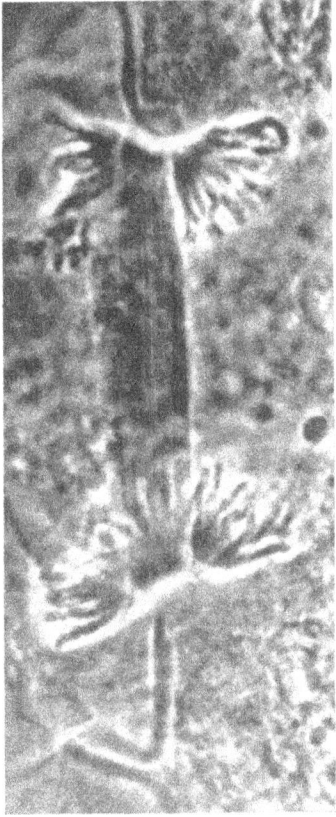

174

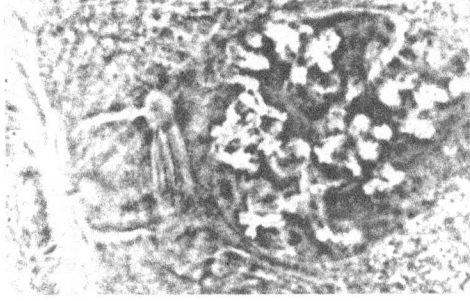

175

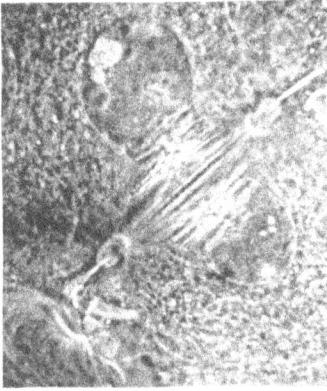

177

176

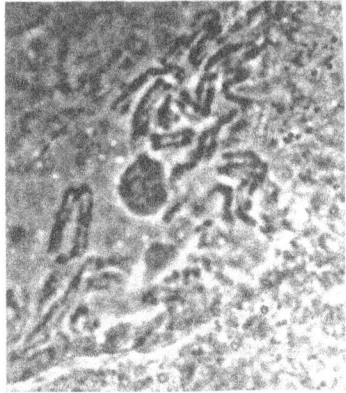

179

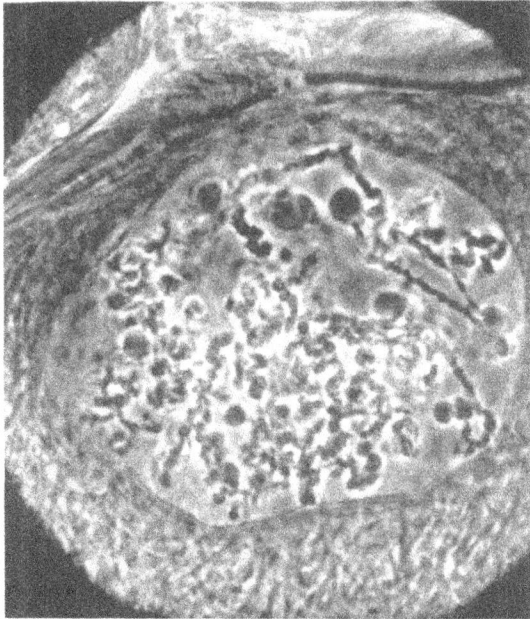

178

865

PLATE 7

FIG. 180. Low power view of anterior end of cell during prophase of meiosis I, showing the two flagellated areas with a
long centriole extending between each area and the end of
the central spindle. The entire nuclear membrane is shown,
but only portions of individual chromosomes are visible.
The long dark bodies on either side of the nucleus are parabasals. × 600.

FIG. 181. Anaphase of meiosis I. Note short, stout chromosomes. Several centromeres are fairly plain. Central spindle is mostly out of focus. × 900.

FIG. 182. Anaphase II. Chromosomes are long and slender.
Several centromeres can be seen (the two near central spindle, upper left, are plain). Central spindle lies in centre of
nucleus. Chromosomes above and below it not in focus;
centrioles also not in focus. × 900.

FIG. 183. The long arms of two late anaphase II chromosomes
(the short arms as well as most of other chromosomes not
in focus). Note large gyres of major coils. One can also
see faintly that the chromonema is surrounded by a matrix.
× 2,000.

FIG. 184. The central spindle portion of the achromatic figure
to show its long strands of fibres. Chromosomes out of
focus and centrioles, except for a short portion of the one
at the upper end, are cut off. × 2,000.

FIG. 185. Late telophase I to show how the two groups of
chromosomes are pushed farther and farther apart as the
rigid, central spindle portion of the achromatic figure increases in length by growth. A few of the centromeres at
either end of central spindle show faintly (they are best at
the lower right end). They connect chromosomes with a
centriole via chromosomal fibres, and the central spindle is
connected, at each end, with the centriole that produced half
of it. But centrioles, in this picture, are out of focus. In
the centre between the two daughter nuclei lies a large
portion of the parent nuclear membrane which is about to
be discarded. Each daughter nucleus is surrounded by its
own small portion of the parent nuclear membrane, which
portion it will later reorganize into a new nuclear membrane. × 600.

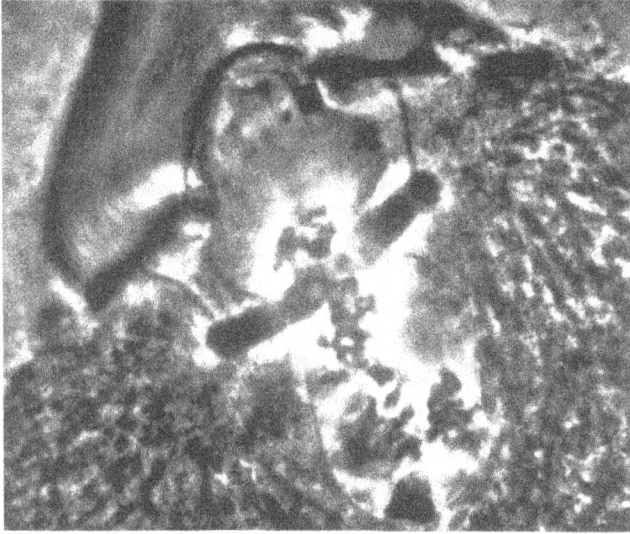

180

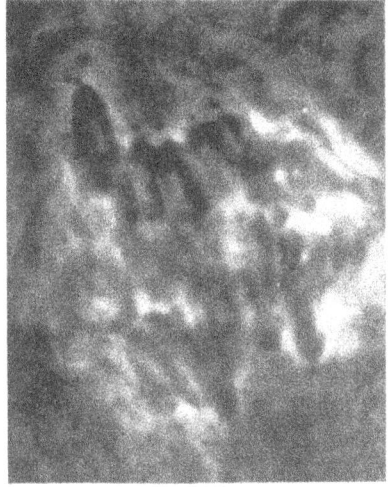

181

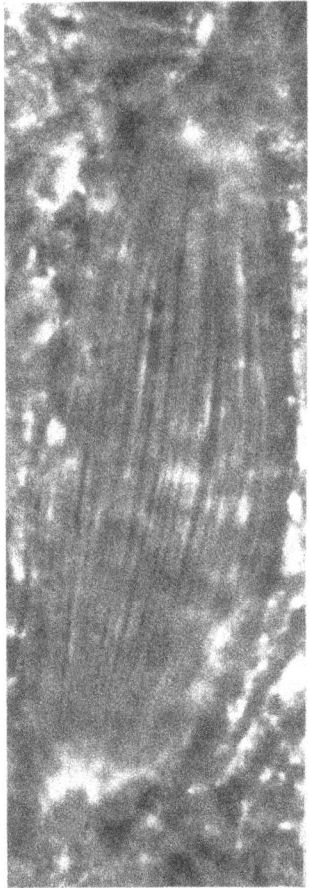

182

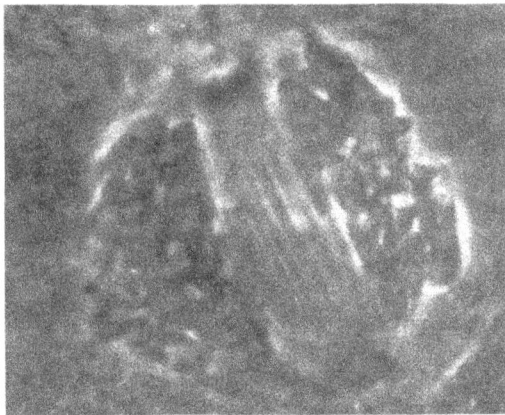

183

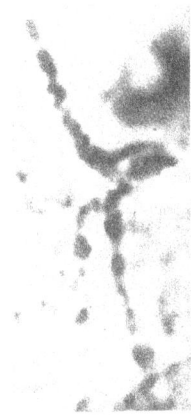

184

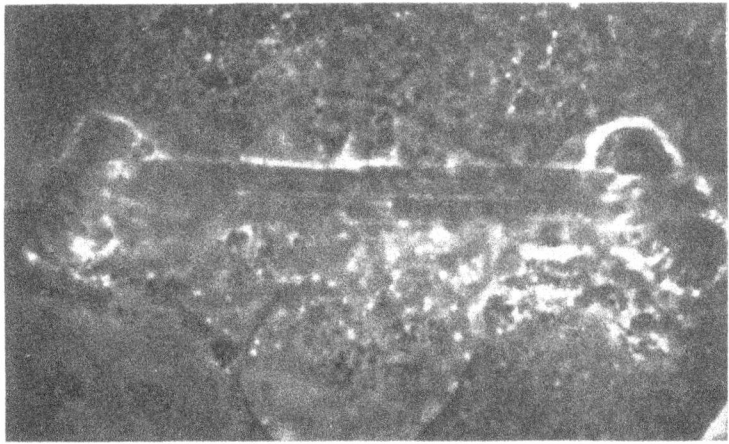

185

PLATE 8

FIG. 186. Anaphase of meiosis II. Note long slender nature of the chromosomes as they go to the poles. Some of the shorter arms of sisters have separated (lost their pairing), but there is still pairing in most of the long arms. Some of the centromeres show fairly well. In several instances one can see that the centromeres of two homologues, for example, are under so much tension that the arms of each are forced to lie side by side, as if they were paired, which, of course, they are not. This is seen best when both arms of a chromosome are of equal length, as in the case of the two homologues at the lower border of the nucleus, where the arms of one are still slightly paired distally to those of the other. In other words, the chromosomal arms of these homologues are forced to lie side by side owing to the simultaneous pull from the centromere and distal ends of each chromosome. Finally, the centromere end wins in this tug of war, as the last vestige of pairing is lost, and the chromosomes become free to move apart rapidly. Note the long central spindle which lies in the centre of the nucleus (the chromosomes below it show faintly; those above are out of focus). There is a long centriole at each end of the central spindle. Chromosomal fibres, which are out of focus in the photograph, connect centriole and centromere. Thus, the elongation of the central spindle pushes the chromosomes apart (compare figs. 185 and 186) and at the same time pulls the intact nuclear membrane in two. × 800.

FIG. 187. Telophase II shortly after cytoplasmic division. The chromosomes are all anchored by their centromeres (and through the centrosome) to the centriole (the large, dark rod near upper left hand corner of the figure). Note the loose major coils which may be seen in many of the chromosomes. Centromeres, at this stage in their life cycle, form almost a complete ring, only half of which can be seen in the photograph. × 1,500.

FIG. 188. Considerably later telophase II. The chromosomes are losing their major coils and thus appear greater in length. The centromeres (just below the large, round centrosome) no longer lie in a ring; they are now in a row. Note portion of the large, elongate centriole in upper left hand margin of photograph. Presently the chromosomal fibres will disappear (dissolve in the cytoplasm) and the connection between chromosomes and centriole will be lost, at which time centriole and nucleus will move apart, not to become connected again until the next cell division occurs. The centromeres always go with the nucleus since they are embedded in the nuclear membrane, but are capable of movement therein. × 1,500.

FIG. 189. Detail of two anaphase centromeres. × 2,000.

FIG. 190. Detail of a tetrad which lies between two free nucleoli (one above to left and the other below to right). All four chromatids are paired in two regions to the left of the lower nucleolus. All four are also paired at each end. But in other places there are regions where only chromatids are paired, particularly near the mid-portion of the tetrad. Just above this mid-region one can see clearly a place where the chromatids exchange pairing partners and thus produce a chiasma. × 2,400.

FIG. 191. Later stage in the shortening and bending (at centromere region) of a tetrad. Beginning at the lower right hand end of tetrad, first there is a region where all four chromatids are paired. This pairing extends to the bend, at which point there is a region of considerable length where the two chromosomes are not paired at all. Then the homologues come together again, and finally, at their upper ends, there is another region where they lie a considerable distance apart. × 2,400.

FIG. 192. A late tetrad lies in the centre. It is greatly bent on itself at the centromere ends, thus forcing the arms of both homologues to lie fairly close together. In the arms to the right, there is complete pairing of homologues; but in those to the left, homologues are paired in some regions and unpaired in others. × 2,400.

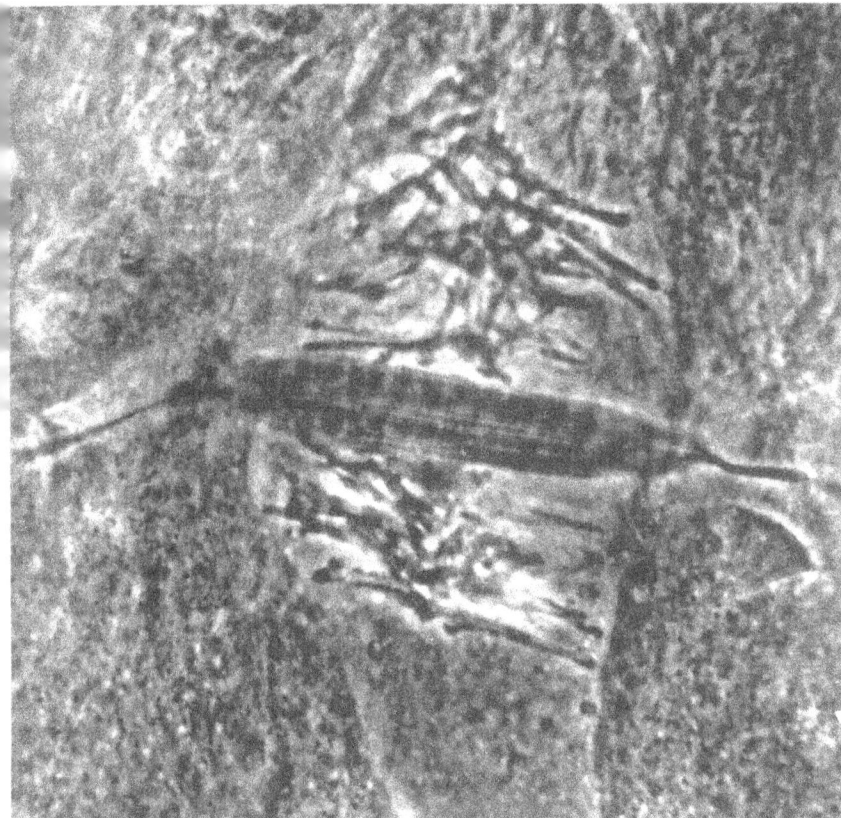

186

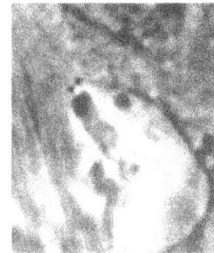

189

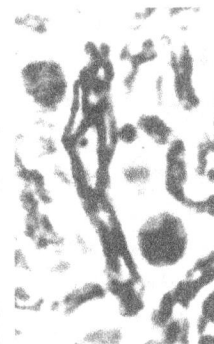

190

191

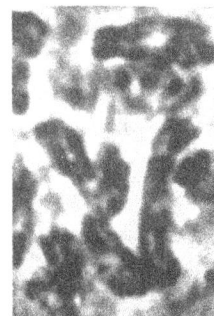

187

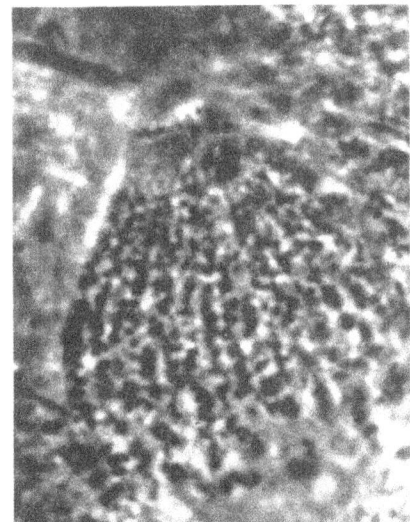

188

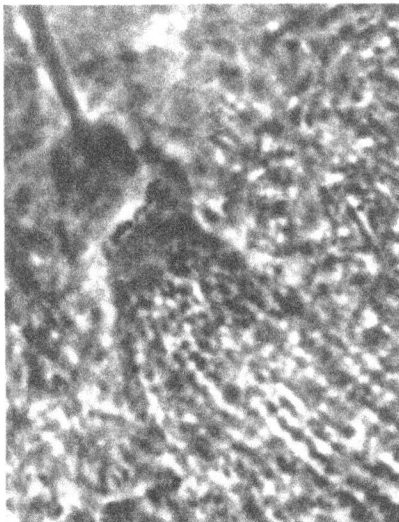

192

www.ingramcontent.com/pod-product-compliance
Lightning Source LLC
Chambersburg PA
CBHW081333190326
41458CB00018B/5987